Janusz Schinke

Processing and Characterization of Organic Thin Films

AF062099

Janusz Schinke

Processing and Characterization of Organic Thin Films

Self-Assembled Monolayers for Interface Engineering and Semiconductors with Thermally Activated Solubility Reduction

Südwestdeutscher Verlag für Hochschulschriften

Impressum / Imprint
Bibliografische Information der Deutschen Nationalbibliothek: Die Deutsche Nationalbibliothek verzeichnet diese Publikation in der Deutschen Nationalbibliografie; detaillierte bibliografische Daten sind im Internet über http://dnb.d-nb.de abrufbar.
Alle in diesem Buch genannten Marken und Produktnamen unterliegen warenzeichen-, marken- oder patentrechtlichem Schutz bzw. sind Warenzeichen oder eingetragene Warenzeichen der jeweiligen Inhaber. Die Wiedergabe von Marken, Produktnamen, Gebrauchsnamen, Handelsnamen, Warenbezeichnungen u.s.w. in diesem Werk berechtigt auch ohne besondere Kennzeichnung nicht zu der Annahme, dass solche Namen im Sinne der Warenzeichen- und Markenschutzgesetzgebung als frei zu betrachten wären und daher von jedermann benutzt werden dürften.

Bibliographic information published by the Deutsche Nationalbibliothek: The Deutsche Nationalbibliothek lists this publication in the Deutsche Nationalbibliografie; detailed bibliographic data are available in the Internet at http://dnb.d-nb.de.
Any brand names and product names mentioned in this book are subject to trademark, brand or patent protection and are trademarks or registered trademarks of their respective holders. The use of brand names, product names, common names, trade names, product descriptions etc. even without a particular marking in this work is in no way to be construed to mean that such names may be regarded as unrestricted in respect of trademark and brand protection legislation and could thus be used by anyone.

Coverbild / Cover image: www.ingimage.com

Verlag / Publisher:
Südwestdeutscher Verlag für Hochschulschriften
ist ein Imprint der / is a trademark of
OmniScriptum GmbH & Co. KG
Heinrich-Böcking-Str. 6-8, 66121 Saarbrücken, Deutschland / Germany
Email: info@svh-verlag.de

Herstellung: siehe letzte Seite /
Printed at: see last page
ISBN: 978-3-8381-5144-1

Zugl. / Approved by: Heidelberg, TU Braunschweig, Diss., 2014

Copyright © 2015 OmniScriptum GmbH & Co. KG
Alle Rechte vorbehalten. / All rights reserved. Saarbrücken 2015

Processing and Characterization of Organic Thin Films: Self-Assembled Monolayers for Interface Engineering and Semiconductors with Thermally Activated Solubility Reduction.

This work deals with two of the major challenges in organic electronics: (i) the optimization of organic/inorganic contacts through interface engineering and (ii) the prevention of intermixing in multilayer solution-processed devices.

(i) A suitable method to control organic/inorganic interfaces, not only in terms of interface energetics but also to control the morphology of subsequent layers is the use of self-assembled monolayers (SAMs). Various SAMs from commercially available precursor molecules have been investigated in detail with the aim to optimize the processing parameters for high-throughput fabrication. On the basis of the gained knowledge and experience SAMs composed of novel molecules with strong dipole moments were investigated. It could be shown that the new SAMs substantially increase the wettability and simultaneously decrease the work function of the used air stable electrodes by more than 1 eV. This is of particular importance since most low work function electrodes are chemically very reactive and oxidize in ambient atmosphere. Finally, the SAMs were deployed in organic field-effect transistors and a significant reduction of the threshold voltage as well as the contact resistance was achieved.

(ii) The second part of this work addresses the problem of intermixing of solution-processed thin films. To this end, the film formation was studied of n-type semiconductor materials based on 1,4,5,8-naphthalenetetracarboxdiimide (NDI) with additional thermally cleavable side groups that govern the solubility of the molecules. Both polymer and small molecule organic semiconductors were studied. In the case of the low-molecular systems fast and uncontrolled crystallization of the layers during pyrolysis resulted in non-uniform films with poor electrical transport properties. The newly synthesized polymers on the other hand provided very encouraging results. The polymers delivered very homogeneous layers which remain almost unchanged after thermal treatment and exhibit excellent solvent resistance. In order to investigate the electrical properties of the semiconductor materials organic field-effect transistors have been prepared and characterized. A mobility of $\mu = 2.4 \cdot 10^{-4}\,\mathrm{cm^2/Vs}$ after pyrolysis was reached.

Prozessierung und Charakterisierung von Organischen Dünnschichten: Grenzflächenoptimierung durch Selbstorganisierende Monolagen und Thermisch Aktivierte Löslichkeitsreduzierung bei Halbleitern.

In dieser Arbeit wurden zwei Themengebiete bearbeitet, die zu den wichtigen Herausforderungen der Organischen Elektronik gehören: (i) die Optimierung von organisch/anorganischen Kontaktflächen mittels Grenzschichtmodifizierungen und (ii) das Verhindern ungewollter Vermischung in flüssigprozessierten Multilagen-Bauteilen.

(i) Eine geeignete Methode um Organik/Anorganik Grenzflächen gezielt zu beeinflussen, sowohl Barrierenhöhen für den Ladungstransport als auch die Benetzbarkeit für weitere Schichten betreffend, ist die Verwendung von selbstorganisierenden Monolagen (SAMs). SAMs aus mehreren kommerziell erhältlichen Ausgangsstoffen wurden hergestellt und auf eine Prozessierung mit hohem Durchsatz hin optimiert. Basierend auf den so gewonnenen Erkenntnissen wurden neuartige SAMs aus Molekülen mit starkem Dipolmoment hergestellt und charakterisiert. Es konnte gezeigt werden, dass die neuen SAMs sowohl die Benetzbarkeit des Substrats erheblich verbessern als auch die Austrittsarbeit um mehr als 1 eV erniedrigen. Dies ist von besonderer Bedeutung, da die meisten verwendeten Elektroden mit einer niedrigen Austrittsarbeit chemisch sehr reaktiv sind und in Umgebungsatmosphäre oxidieren. Anschließend wurden die SAMs in organischen Feld-Effekt-Transistoren eingesetzt, wo sie zu einer deutlichen Senkung der Schwellenspannung als auch des Kontaktwiderstandes führen.

(ii) Im zweiten Teil der Arbeit wurde die Problematik der Vermischung von flüssig-prozessierten Dünnschichten bearbeitet. Neue n-Typ-Halbleitermaterialien auf Basis von 1,4,5,8-Naphthalen-tetracarbonsäurediimid (NDI) mit löslichkeitsvermittelnden Seitengruppen, die bei moderaten Temperaturen thermisch abspaltbar sind, wurden untersucht. Hierbei kamen sowohl kleine Moleküle als auch Polymere zum Einsatz. Bei den kleinen Molekülen führte eine schnelle und unkontrollierte Kristallisation während der Pyrolyse zu ungleichmäßigen Schichten mit schlechten elektronischen Transporteigenschaften. Die Polymer-basierten Schichten dagegen blieben auch nach der Pyrolyse homogen und zeigten zudem eine hervorragende Stabilität in Lösemitteln. Auf diese Weise hergestellte Feld-Effekt-Transistoren erreichten Mobilitäten von $\mu = 2.4 \cdot 10^{-4}$ cm^2/Vs nach der Pyrolyse.

Contents

1. **Introduction**   1
   1.1. Scope of this Work   2
   1.2. Outline   4
2. **Theoretical Background**   7
   2.1. Organic Semiconductors   7
      2.1.1. Charge Transport   8
      2.1.2. Interfaces   10
   2.2. Self-Assembled Monolayers (SAMs)   12
      2.2.1. Structure and Morphology   13
      2.2.2. Growth   14
      2.2.3. Electrical Properties   17
3. **Experimental Details**   21
   3.1. Methods   21
      3.1.1. Kelvin Probe   21
      3.1.2. Photoemission Spectroscopy   22
      3.1.3. Infrared Spectroscopy   26
         3.1.3.1. Transmission of Light in Three-layer Systems   29
         3.1.3.2. FTIR-Spectroscopy   32
         3.1.3.3. Infrared Reflection-Absorption Spectroscopy - IRRAS   33
      3.1.4. Atomic Force Microscopy   35
   3.2. Sample Preparation   37
      3.2.1. Preparation of SAMs   37
         3.2.1.1. Gold Substrates Cleaning Procedure Investigation   38
         3.2.1.2. ITO Substrates   44

Contents

		3.2.1.3. Treatment	44
	3.2.2.	Organic Semiconductors with thermally activated solubility reduction	47
		3.2.2.1. Substrate Preparation	47
		3.2.2.2. Pyrolysis	48
3.3.	Devices		48
	3.3.1.	Organic Solar Cells	49
		3.3.1.1. Working Principle	49
		3.3.1.2. Organic Solar Cells - Design and Stack Used in this Work	50
	3.3.2.	Organic Field Effect Transistors	51
		3.3.2.1. Working Principle	51
		3.3.2.2. Organic Field Effect Transistors - Stack and Layout	53
	3.3.3.	The Clustertool	53

4. Polymeric Charge Injection Layers — 59

4.1.	State of the art	59
4.2.	PCILs on Various Substrates	61
	4.2.1. PCILs on Metals - Characterization	61
	4.2.2. PCILs on ITO - Characterization	63
4.3.	Devices	66
4.4.	Summary	69

5. Self-Assembled Monolayers — 71

5.1.	State of the art	71
5.2.	Preliminary Examination	73
5.3.	Perfluorodecanethiol (PFDT) as Model System	76
	5.3.1. PFDT - PES characterization	76
	5.3.2. Exposure Dependent Monolayer Formation and Work Function Shift	78
	5.3.3. Discussion	85

	5.4.	Novel SAMs	88
		5.4.1. Aromatic Thiols on Gold	89
		5.4.1.1. Properties and Different Growth Mechanisms of the New Molecules - PES and IR	92
		5.4.1.2. OFET Devices	98
		5.4.2. Aromatic Phosphonic Acids on ITO	100
	5.5.	Conclusion	102

6. Organic Semiconductors with thermally activated solubility reduction   105

	6.1.	State of the art	106
		6.1.1. Small Molecules	107
		6.1.2. Polymers	108
	6.2.	NDI Derivatives - Investigation of Synthesized Small Molecules	108
		6.2.1. Dewetting and Pyrolysis	109
		6.2.2. Summary	113
	6.3.	NDI Derivatives - Investigation of Synthesized Polymers	114
		6.3.1. P(HtHC-NDI-4HT2) → P(NDI-C6OH-4HT2) Molecular Weight of **130 kDa**	115
		6.3.1.1. Morphology	116
		6.3.1.2. OFETs	117
		6.3.2. P(HtHC-NDI-4HT2) → P(NDI-C6OH-4HT2) Molecular Weight of **200 kDa**	119
		6.3.2.1. Morphology	119
		6.3.2.2. XPS and IR results	120
		6.3.2.3. OFETs	126
		6.3.3. P(tHC-NDI-4HT2) → P(NDI-4HT2)	127
		6.3.3.1. Morphology	128
		6.3.3.2. XPS and IR	129
		6.3.3.3. OFETs	132
		6.3.4. P(HtODC-NDI-T2) → P(NDI-C6OH-T2)	133
		6.3.4.1. Morphology	135
		6.3.4.2. XPS and IR	136
		6.3.4.3. OFETs	140
		6.3.5. Comparison and Discussion	142

Contents

    6.4. Conclusion 144

7. Summary   **147**

Bibliography   **150**

A. Appendix   **177**

Journal Publications, Patents, Conference Presentations and Supervised Theses   **185**

1. Introduction

Shortly after the invention of the inorganic bipolar transistor by W. Schockley, J. Bardeen and W. Brattain in the late 40's [1] microelectronics became an integral part of every days lives and are indispensable in modern information society. One drawback of modern semiconductor technology is the production process of monocrystalline low-defect silicon, which requires complex and expensive epitaxy and cleaning procedures. In recent years there has been growing interest in the field of organic semiconducting materials. Already back in 1906 Pochettino reported about photoconductive properties of anthracene [2] and in 1962 Pope et al. demonstrated electroluminescent properties of the same material [3]. One of the breakthroughs was delayed until 1977 and came first with the discovery of conducting polymers based on π conjugated molecular systems done by A. J. Heeger, A. G. MacDiarmid and H.Shirakawa [4]. The first successful application of π conjugated molecules for the preparation of first organic electroluminescent diodes was reported by Tang and VanSlyke back in 1987 [5]. Organic materials offer the advantage of deposition at relatively low temperatures from the gas phase on various substrates like transparent indium tin oxide (ITO), and even printing on flexible substrates is possible [6]. This enables a wide range of innovative products such as transparent and flexible displays, solar cells or in general flexible electronic circuits in large format, that cannot be realized using inorganic materials, or only at very high costs. Due to the first successes described above the research efforts in the last three decades have been greatly enhanced and today there are already numerous products on the market, whose function is based on organic semiconductors. The most common applications so far are displays based on organic light emitting diodes (OLEDs) which due to their low power consumption are widely used especially in mobile devices such as smartphones and tablets, but the first affordable OLED TVs are also already on the market. One of the biggest remaining problems other devices

1. Introduction

based on organic semiconductors are still the low lifetime and sometimes the insufficient efficiency of the components, especially compared to inorganic components. Due to intrinsic limitations organic electronics will most probably never replace the silicon-based technology, but this is not necessarily the goal of the organic semiconductors. Much more important is the ability to add completely new applications or merge with the already existing inorganic technologies. Therefore, even more research effort is needed to exploit the potential of organic semiconductors in corresponding components.

1.1. Scope of this Work

In 2008 the cluster Forum Organic Electronics was founded by the Federal Ministry for Education and Research (BMBF), which combines the research activities of numerous universities and currently 30 companies in the field of organic electronics to promote the development of this future technology. The main focus is to integrate the entire value chain from design and synthesis of new materials, to the study of their fundamental properties and at the end to develop the components and suitable manufacturing processes of the products and their applications. The heart of this project is located in Heidelberg. InnovationLab GmbH [7] is a research and transfer platform with over 100 employees and high-tech laboratory infrastructure, which allows to jointly research on fundamental properties as well as to develop new innovative products . The present work was carried out within the scope of the BMBF-project MORPHEUS[1], which deals with development and characterization of new injection layers and new organic semiconductors that address the fundamental problems in organic electronic devices.

The MORPHEUS project, in its two parts, concentrates on two major topics which still cause issues or hinder the preparation process of various multilayer organic devices. A great challenge for printed organic electronics is to deposit the subsequent layer of a multilayer component from the same solvent without destroying the underlying freshly deposited layer (see Figure 1.1). One way to overcome this problem is to develop materials whose solubility can be specifically influenced by an external stimulus. Such reactions do not require the addition of a reagent and have the advantage that eventually only

---

[1] Morphologiekontrolle für Effiziente und Stabile Bauelemente.

1.1. Scope of this Work

minor amounts of impurities remain in the deposited film. Thus only little possibility remains for the impurities to interfere with the semiconducting properties or any desired morphological changes within the active layer of the used organic material. Since many solution processed organic semiconductors require a thermal treatment after deposition to adjust the morphology, it makes sense to integrate the solubility reduction step in this processing stage and resort to thermally induced reactions. It is possible to provide solubilizing groups by a labile unit and cleave them after deposition by a specific external stimulus. Ideally, all these cleavage products are volatile and do not remain after pyrolysis in the film, and so cannot disrupt the electronic properties of the semiconductor. A detailed description of the current development in thermal cleavage of solubilizing side chains in organic semiconductors and their application can be found in the review article by Krebs et al. [8]. In this work several new synthesized 1,4,5,8-naphthalenetetracarboxdiimide (NDI) derivatives[2] which act as n-type semiconductors were investigated on their solubility and processability as a thin film. The main focus was on low pyrolysis temperature, high solvent resistance and good semiconducting properties.

The next major challenge in the organic electronic devices and simultaneously in the second part of MORPHEUS project was the optimization of interfaces in multilayer structures. The interfaces between conducting and semiconducting materials play an important role. The aim of the project is to control barriers for charge carriers at these interfaces in order to increase the device performance. There are two main approaches: either by influencing the morphology of the semiconducting material or by adjusting the electronic properties. Since the electronic properties of organic semiconducting materials are difficult to manipulate, a great deal of work is concentrated on adjusting the electronic properties of air-stable electrode materials like noble metals or widely used transparent oxides. A promising way to do so is to apply a self assembled monolayer (SAM) with an intrinsic dipole onto the electrode surface altering its surface potential and therefore its macroscopic electronic properties (see Figure 1.2). The aim of the work is to create compounds which are capable to influence wettability and electronic aspects of the used substrates as well as to control the morphology of subsequent layers. The new SAM

---

[2] Synthesized by Torben Adermann from University of Heidelberg, OCI.

1. Introduction

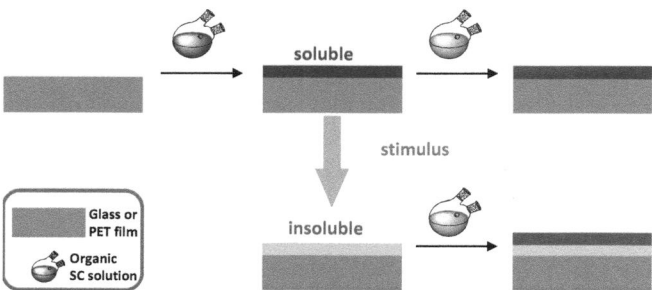

Figure 1.1: Schematic depiction of mixing problems (top) by solution processing of advanced layer sequences. Bottom: alternative route with modified semiconductor materials which should change their solubility upon an external stimulus after being deposited from solution.

molecules were designed and synthesized by the group of Manuel Hamburger (OCI, University of Heidelberg). These compounds resemble commonly used organic semiconducting materials and concurrently contain an appropriate dipole moment in order to alter the work function of the electrode. In this work several monolayers were prepared and investigated by different analytical methods as well as in organic field effect transistors. Studies towards the formation of monolayers on various substrates using these molecules and their influence on the electronic properties will be shown in this thesis. Furthermore, preliminary device testing will be presented.

1.2. Outline

The present work is organized as follows. Section 2 briefly describes the theoretical background necessary to understand the results. In Chapter 3 all characterization methods used throughout this work are described in detail. Next the sample preparation processes are depicted and an extensive substrate cleaning procedure is introduced which proved to be necessary to achieve high quality monolayers. Following organic electronic devices used to characterize investigated semiconductors or self-assembled monolayers are introduced. Finally, the integrated UHV system is described where several

1.2. Outline

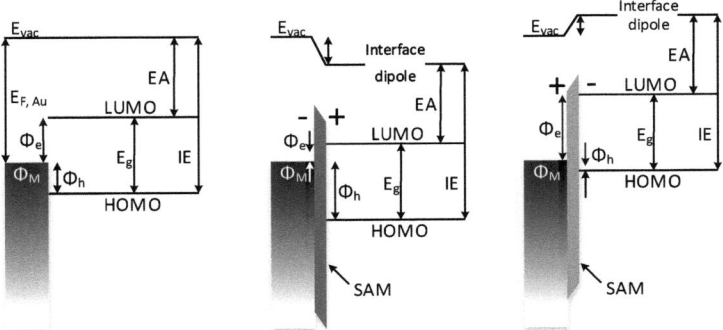

Figure 1.2: Schematic representation of energy level alignment of metal/organic interface with metal work function ϕ_M within the energy gap E_g. Left: without injection layer, middle: electrode treated with SAM which lowers the work function and decreases the electron injection barrier, right: analog but with SAM which increases the work function of the used electrode.

devices were prepared. Chapter 4 shows experimental data of polymeric charge injection layers (PCILs), which became benchmark material for lowering the work function of different electrodes. Some new insights could be obtained but the possibility of using these materials as reference with respect to materials introduced within this work was even more important and especially challenging. Self-assembled monolayers are introduced in Chapter 5. First the preparation process on commercially available molecules is established and the possibility of high-throughput processes like printing is investigated. Newly synthesized molecules were used as monolayers and investigated with various analytical methods as well as used in OFET devices. Section 6 shows the results of the second part of the MORPHEUS project, namely organic semiconductors with thermally activated solubility reduction. At first new small molecules are introduced and results are presented. The second part of this section reports on the results obtained from new synthesized polymers with the same thermally activated solubility reduction feature. Various analytical techniques like atomic force microscopy, photoemission spectroscopy

1. Introduction

and infrared spectroscopy were used to characterize the organic semiconductor layers. In order to prove their functionality after the pyrolysis process OFET devices were prepared and characterized. Finally Chapter 7 concludes the thesis with a short summary and gives a brief outlook on possible future work.

2. Theoretical Background

The purpose of this chapter is to introduce the topic of organic electronics and provide a knowledge base that is helpful to understand the upcoming chapters. In the first section an introduction to organic semiconductors will be given. The theoretical background on self-assembled monolayers is presented in the second section. Finally, basic theory on charge injection at contacts will be depicted.

2.1. Organic Semiconductors

Almost all compounds containing carbon atoms are described as organic materials[1] [9]. In organic semiconductors carbon atoms are bound with so called conjugated double bonds, which means that carbon atoms are held in place by the alternation of single and double bonds. This is in contradiction to the well known valence bond theory [10, 11]. Carbon has two valence p-electrons and according to the above mentioned valence bond theory only two bonds to neighboring carbon atoms are allowed. In order to explain the carbon structure of organic semiconductors, the concept of hybridization must also be introduced. *Hybridization is the conceptual mixing of atomic orbitals in order to form new hybrid orbitals which are suitable for the qualitative description of atomic bonding properties* [12, 13].

In the valence shell of a carbon atom, there are four electrons. The 2s-orbital is fully occupied by two electrons and two 2p-orbitals are partially filled each by one electron. If a carbon atom bounds to another carbon atom, the outer atomic orbitals of each atom are able to form combined hybrid orbitals in order to reach an energetically favorable state. Depending on the bonding type three different types of hybridization are possible. For the electronic properties of

---

[1] Simple carbon-based compounds like for example CO and CO_2 are classified as inorganic.

2. Theoretical Background

organic semiconductors, the formation of delocalized π electron system, which is formed by sp^2 hybridized carbon atoms, is of fundamental importance. In this type of hybridization, the s-orbital and two p-orbitals are forming three sp^2 hybridized orbitals each occupied by one electron [14, 15]. None of the three sp^2 orbitals is energetically favored and the angle between each bond is the same (120°). The remaining fourth electron is located in the not hybridized p-orbital and can form so called π bonds perpendicular to the plane of the sp^2 orbitals. The electrons in these π orbitals contribute to the electrical conductivity. Thus, the model of hybridization can qualitatively explain the conjugated double bonds of carbon atoms in organic semiconductors.

If two carbon atoms bind in the sp^2 hybridization manner, a bonding σ orbital and an antibonding $\sigma *$ orbital is formed. They are strongly split and localized due to the strong overlap of the two orbitals. In the same way a bonding π orbital and an antibonding $\pi *$ orbital are formed. However, these are less split due to the lower spatial overlap. σ and π form a double bond between the two carbon atoms. The energy levels of two carbon atoms in the sp^2 hybridization and the σ and π bonds between two carbon atoms are schematically depicted in Figure 2.1.

In molecules whose structure is dominated by carbon atoms, the sp^2 orbitals form also a σ bond between the carbon atoms. The overlap of the p_z orbitals leads to the formation of bonding π and antibonding $\pi *$ orbitals. In the ground state, all states of the π orbital are occupied and it forms the highest occupied molecular orbital (HOMO). The antibonding $\pi *$ orbital is the lowest unoccupied molecular orbital (LUMO). In Figure 2.2 a benzene ring consisting of several carbon atoms in the sp^2 hybridization is presented. The electrons in the HOMO are delocalized over the entire molecule (blue) due to the delocalized π electron system and cannot be assigned to a single carbon atom.

2.1.1. Charge Transport

In solids or thin layers consisting of organic semiconductors the molecules are only weakly bound to each other (van der Waals forces). Due to the low intermolecular overlap of the π orbitals, the charge carriers are localized on individual molecules. Therefore the charge transport is usually described as hopping transport [14] of the charge carriers from one molecule to the next

2.1. Organic Semiconductors

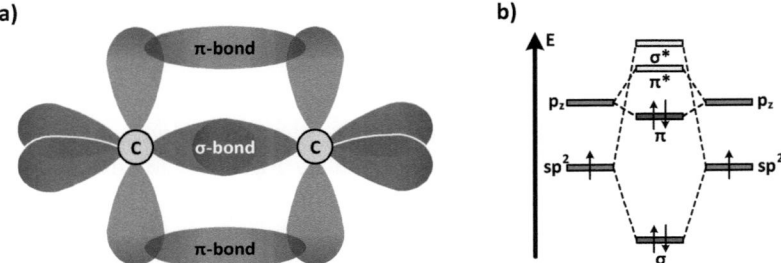

Figure 2.1: a): Schematic representation of molecular orbitals of C-C bond. b): Corresponding representation of the energy levels in the sp^2 hybridization as well as the bonding/antibonding orbitals.

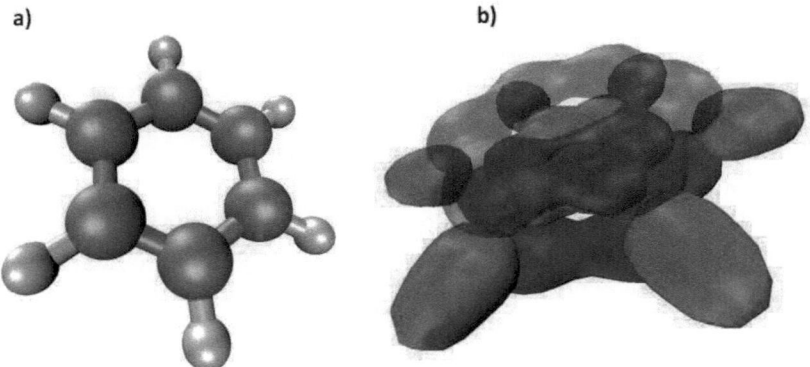

Figure 2.2: a): 3d drawing of a benzene ring. b): Corresponding 3d representation of the hybrid orbitals of benzene with σ (red) and π (blue) bonds and the delocalized π-electron system.

2. Theoretical Background

one. The hopping transport process requires thermal activation to overcome energy differences [14]. The energy needed for the hopping transport is called the activation energy E_A, which is the difference between the mean energy of charge carriers $\overline{E}$ and the transport energy E_T:

$$E_A = \overline{E} - E_T. \qquad (2.1)$$

This leads to low mobilities μ of about 10^{-3} cm²/Vs to 10^{-5} cm²/Vs in amorphous organic semiconductors [16]. Even in crystalline organic semiconductors, the mobility at room temperature is in the range of 1 cm²/Vs [17] and thus several orders of magnitude lower than the mobility of inorganic semiconductors (for crystalline silicon $\mu \approx 1340$ cm²/Vs [18]). A comprehensive study about the charge transfer in organic semiconductors, in particular hopping transport is presented in [14, 19]. No matter if inorganic or organic semiconductors, in both cases the charge transport usually works better in ordered systems. Additional barriers for the charge transport are caused by disorder or grain boundaries for example. Another source of barriers for charge carriers in an electronic device are the mismatched contacts and interfaces.

2.1.2. Interfaces

Most organic electronic devices consist of at least two different layers: the electrode and the active organic material. More often, however, even more complicated device structure with different specialized organic layers is used. Therefore the organic-metal and organic-organic interfaces play crucial roles in charge injection or/and in charge transport in the devices. In order to produce efficient and stable devices, and to reach similar control level of their behavior, like in case of inorganic semiconductor devices, the electronic structure, chemical properties and electrical behavior of organic materials and their interafces need to be fully understood. At the beginning of the intensive studies on organic electronic devices, it was assumed that the electronic structure of metal-organic interfaces would follow the basic rule of vacuum level alignment known as the Schottky-Mott limit (see Figure 2.3a) even though its validity at inorganic semiconductor interfaces was already disproved [20]. According to this concept, a prediction of charge injection barriers at the metal-

2.1. Organic Semiconductors

organic interface can be made. The hole injection barrier ϕ_h is defined as the difference between ionization energy of the organic material (IE) and the metal work function ϕ_M. The difference between ϕ_M and the electron affinity (EA) of the organic layer describes the electron injection barrier ϕ_e. However, in reality the organic interfaces are far more complex [21, 22]. Fermi level pinning, which is a well known phenomenon in inorganic semiconductors [23, 24], and formation of interface dipoles must be considered [25–27]. In Figure 2.3b, a more complex picture of the energy alignment of the metal-organic interface is presented. Again ϕ_h and ϕ_e show the hole and electron barriers respectively. $E_\text{vac(M)}$ and $E_\text{vac(O)}$ indicate the metal and organic vacuum level which change due to the formation of interface dipole (Δ Dipole barrier). The values of ϕ_h and ϕ_e are also affected by the interface dipole. Moreover, the morphology, which is strongly influenced by different processing approaches, plays an important role for the interface properties in organic electronics. Even in which order the metal-organic interface is formed has to be considered. Due to the softness of organic materials, the deposition sequence has a strong impact on the interface, especially if vacuum evaporation is used. In [20] a photoemission spectroscopy investigation is performed on interfaces between various metals and eight different organic materials. The results of this experiment show how important direct investigations and measurements of interfaces are and most importantly how poor the predictions of the Schottky-Mott model work in reality.

In summary, a prediction of the interface between different materials according to the energy levels of single materials is very difficult and very often incorrect. For this reason, direct investigations and measurements on interfaces are of great importance. In the PhD thesis off Rebecca Saive [28], a measurement technique is presented, which allows to measure the interface not only on horizontal devices but also on vertical multilayer devices like OSCs. However, the measurements are only one side of the coin. The other one, at least as important, is the way to directly adjust the properties of the interface.

2. Theoretical Background

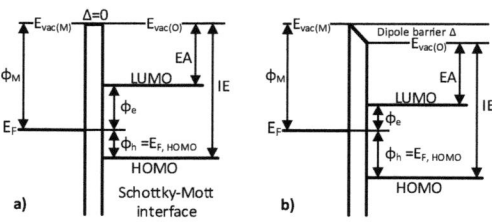

Figure 2.3: Energy diagrams of metal organic interface with (b) and without (a) a dipole barrier. Taken from [20].

2.2. Self-Assembled Monolayers (SAMs)

Self-assembled monolayers (SAMs) are important means for the functionalization of surfaces at the nanoscale and offer an extremely interesting field for the optimization of interfaces in organic electronics [29]. The monolayers are created by adsorption and spontaneous organization of amphiphilic surface-active substances either from solution or by deposition from a gas phase. The substances must ensure an appropriate anchor group like thiols (R − SH) or phosphonic acids (C − PO(OH)$_2$ or C − PO(OR)$_2$) for different surfaces. A well known example is the treatment of the gate insulating SiO$_2$ layer in organic field effect transistors with n-octadecyltrichlorosilane (OTS, silanes have Si$_n$H$_{2n}$ as anchor group). It has been shown that the SAM surface modification improved the crystal structure of the organic semiconductor TIPS-pentacene, so that the charge carrier mobility μ was increased noticeably [30]. In addition, the number of trap states was reduced at the interface between the organic semiconductor and gate dielectric by the octadecyltrichlorosilane layer, which is again beneficial for the charge carrier mobility [31]. However, SAMs are not only limited to having a significant impact on the morphology of the thereon growing structures but they also show a strong influence on the electrical properties of the treated surface. In this work, the focus is mainly set on, but not limited to, the characterization of metal surfaces (especially gold) with thiol-based SAMs. Particularly, gold is suitable as a substrate for SAMs,

2.2. Self-Assembled Monolayers (SAMs)

because it does not oxidize under atmospheric conditions and shows only a little tendency to react with other chemicals. In the next sections, the structure of SAMs and their growth mechanisms will be described as well as the influence on electrical properties.

2.2.1. Structure and Morphology

SAMs often show a high degree of order. Figure 2.4 schematically shows the basic structure of a single adsorbed SAM molecule. By chemisorption of the anchor group to the substrate surface, even strong covalent bonds are possible (in case of thiol-based SAMs only a semi-covalent bond is created $\sim 100\,\mathrm{kJ/mol}$ [32]). As long as the bond to the surface is ensured by the anchor group, any constituent can be freely substituted. This allows a wide range of functionalization by varying the chemical composition. The SAMs usually do not adsorb orthogonally to the surface, but are slightly tilted at an angle θ to the surface normal, which is characteristic for the used substance and substrate. The backbones (in Figure 2.4 marked as alkyl backbone) of the molecules interact through the van der Waals interaction with the same adsorbates, so that densely packed and well-structured monolayers with layer thicknesses between $0.5 \rightarrow 5\,\mathrm{nm}$ organize spontaneously [29, 33]. In addition to the tilt angle, some individual adsorbates show a rotation angle β. Using various functional head groups, the interfacial properties as well as the layer growth can be easily controlled. Depending on the used functional head group, a controlled layer-by-layer growth [34–36] or a self-terminating arrangement to exactly one monolayer [37] with very precise layer thickness and tilt angle to the surface control is possible. Moreover, the head group can have an immense impact on the surface energy which is important for processing conditions and of course an adjustment of the work function (WF) is also possible. In organic electronics, the surfaces which are considered for SAMs treatment are limited to electrodes (mostly noble metals but also some conductive oxides like ITO), dielectrics and organic semiconductors [38].

2. Theoretical Background

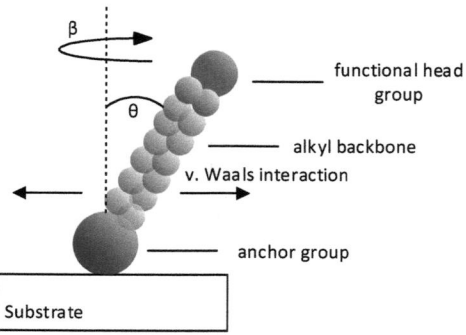

Figure 2.4: Schematic representation of a simple SAM molecule.

2.2.2. Growth

The adsorption is trying, like each thermodynamic system, to achieve its lowest energy state. Figure 2.5 shows the route of the free molecule in solution up to the fulfilled chemisorption. The process is explained on the example of a thiol molecule and a gold surface (other molecules behave analogously, with some differences in binding reaction). A molecule which approaches the substrate (diffusion) loses thermal energy due to the inelastic collision with the surface. If the energy loss is too large, the molecule will form a dipole-dipole interaction with the surface atoms due to the statistical fluctuations of the charge carriers and thus physically adsorb in the Lennard-Jones potential (right potential well). This step is referred to as the "lying-down" phase, because the molecules are initially placed without any order on the interface. The molar enthalpy of physisorption ΔH_P of n-alkanethiols for short carbon chains is in the range of $50 \rightarrow 100$ kJ/mol [39]. The physisorbed molecules, also known as precursors, which are able to additionally provide the amount of activation enthalpy ΔH_PC will undergo the transition into the Morse-potential. There the molecules change over to an excited vibronic state and relax in the ground state. The disexcitation takes place due to the diffusion and the heat exchange at the surface. The sulfur atom of the thiol anchor group

2.2. Self-Assembled Monolayers (SAMs)

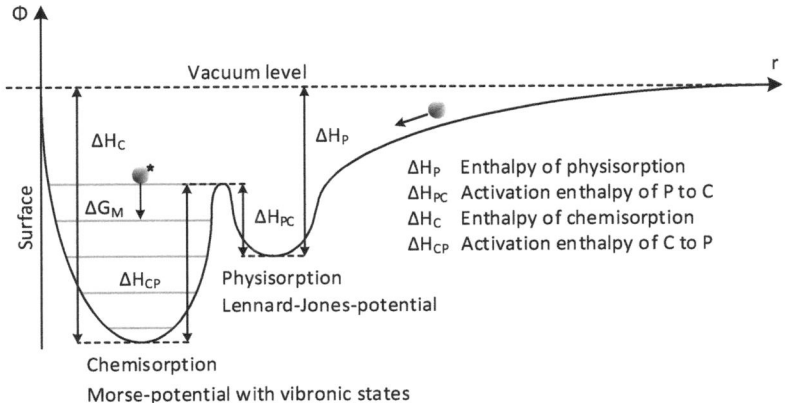

Figure 2.5: Schematic representation of the path of a molecule (ball) through the Lennard-Jones potential to the ground state of the Morse potential via vibronic states. The barriers between physisorption and chemisorption are referred to activation enthalpies ΔH. The free energy ΔG_M shows the direction of the driving force of the adsorption [44].

form in this case a semi-covalent bond with a gold atom [40]. The proton of the hydrogen atom is dissociated homolytically [41]. A negatively charged thiolate-gold bond ($Au - S^-$) remains on the surface. However, the fate of the hydrogen, which most probably desorbs from the surface as H_2, is until now not proved [42, 43]. Chemisorption results in a change in the chemical structure, whereas the physisorption does not. The molar enthalpy of chemisorption was determined for n-alkanethiols as 128 kJ/mol independently from the chain length [39]. Despite the semi-covalent bond, it is still quite possible that molecules overcome the barrier from the excited vibrational state and come back to physisorption or even desorb completely.

The process can also be macroscopically described by a thermodynamic potential G as a function of the state variables pressure p, temperature T and molar solution of n. The relationship is given by the Gibbs equation:

$$\Delta G = \Delta H - T\Delta S. \tag{2.2}$$

2. Theoretical Background

The G is also known as the Gibbs free enthalpy and is used as a guide value for the driving force of the adsorption process [45]. Therefore, a negative Gibbs free energy points always in the direction of the minimum and thus the required equilibrium. SAMs offer the advantage that they react exergonic (voluntarily, $\Delta G < 0$) and require no external energy sources. Therefore, the chemisorption must be an exothermic ($\Delta H < 0$) and thus under steady increase of entropy an irreversible reaction [46]. The ensemble of thiol molecules will attempt to reach the thermodynamic equilibrium with the surface during adsorption. The balance is achieved when the above potential reaches its minimum ($\Delta S = 0$). In other words, as many molecules as possible must firmly anchor in the minimum of the Morse-potential. If the surface is divided into energetically equivalent adsorption positions, each anchor corresponds to the occupation of exactly one such position. A thermodynamic equilibrium ($\Delta S = 0$) also implies, in addition to the thermal and electronic equilibrium, a chemical one, which is dynamic and based on a constant ratio between adsorption (k_+) and desorption (k_-) [47]. In this state, the equilibrium constant K of the reaction is defined appropriate to the law of mass action by $K = \frac{k_+}{k_-} = e^{-\frac{\Delta G°}{RT}}$. $\Delta G°$ is the free standard enthalpy that is substance specific and can be found in the literature. The degree of coverage Θ of the surface with the equilibrium constant K is described by the Langmuir isotherm [48]:

$$\Theta = \frac{KC}{1 + KC}. \qquad (2.3)$$

The degree of coverage is plotted as a function of the solution concentration C for different constant temperatures (see Figure 2.6 left). The curve shows that the coverage rises very quickly, but for higher concentrations goes slowly into saturation. Therefore it does not make much sense to increase the molecule concentration in the solution in order to achieve denser monolayers. The rate of deposition is often compared to the Avrami model $\Theta(t) = 1 - \exp(-kt^n)$ and is well described in [33]. The exponent n refers to the so called Avrami exponent which depends on the degree of crystallization. The rate constant k is described by the temperature and the activation enthalpy in the Arrhenius form. In Figure 2.6, the curve on the right plot shows why several hours of immersion time are needed to achieve better results in the SAM treatment. The "lying-down" phase is mostly completed within a few minutes. The

2.2. Self-Assembled Monolayers (SAMs)

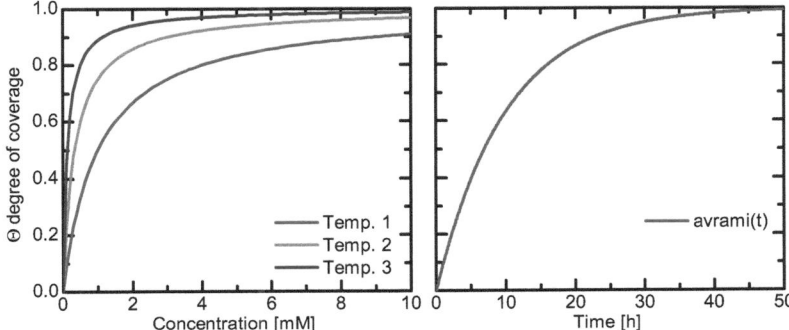

Figure 2.6: Left: Langmuir isotherms show the degree of coverage as a function of the solution concentration at different temperatures. Right: Avrami model of the degree of coverage as a function of time.

organization of densely packed and highly organized monolayers is a slow process due to the van der Waals interaction between the backbones of the SAM molecules [49].

The observed growth phases of the monolayer are sketched in Figure 2.7. The arrangement may be seen chronologically. The chemisorbed molecules which are in an excited vibrational state are able to switch back and forth between the adsorption positions until a suitable partner molecule for the van der Waals interaction is found, which allows to straighten up and thereby builds the first nucleus (nucleation phase). The nuclei will continue to build up "islands" (organization phase) until the whole surface is covered and found its favorable energy state [33].

2.2.3. Electrical Properties

Interfaces between semiconductors and electrode contacts often imply a barrier that inhibits charge carrier injection or extraction (see Section 2.1.2). SAM molecules can be adjusted in their electronic properties such that they are able to purposefully minimize this barrier [50–52]. In physisorbed or chemisorbed systems, a rearrangement of the charge carrier distribution takes place, resulting in the formation of an interface dipole, which can be used specifically by

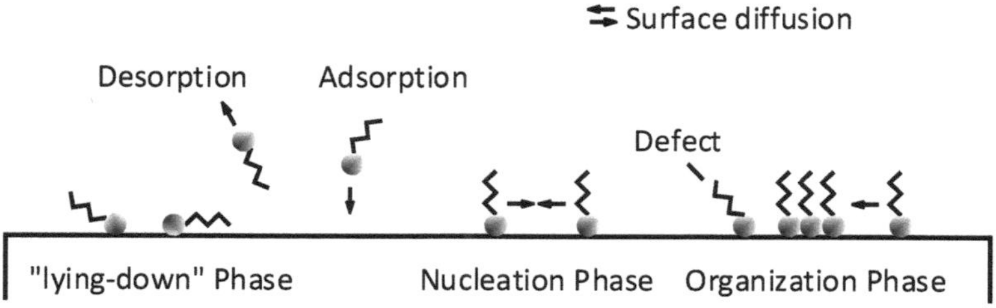

2. Theoretical Background

Figure 2.7: Chronological representation of the experimentally observed growth phases.

means of SAMs to change the WF of the given electrode. The chemisorption of SAMs is dissociative, which means chemical bonds are broken and new ones are established. In case of thiol-based SAMs and gold substrate the free thiol group (S − H) is replaced during the adsorption by a negatively charged thiolate (Au − S$^-$), resulting in a first bond dipole [53]. Concurrently, the wave functions of both materials are overlapping. Due to the Pauli principle, the exponentially outgoing electron density of states is compressed on the surface of the used metal. This is often referred to the so called "pillow" or "push back" effect [54]. Both effects can be summarized to a bond dipole Δ_{BD}. Additionally, the molecule has a permanent dipole moment μ_0 that is composed of the vector sum of all polar bond dipoles of the used constituents [55]. However, the interaction of all the dipole fields in a monolayer results in a depolarization of the individual molecules. The charge carrier fluctuations within the molecules, which are especially pronounced in the area of the functional head group can be associated with the depolarization [56]. Thus, the SAM molecule has different ionization potentials between anchor and head group. The difference in the ionization potentials is expressed by a shift of the vacuum level ΔE_{vac} as:

$$\Delta E_{\text{vac}} = \frac{e\mu_\perp}{\epsilon_0 A} \rightarrow \frac{e\mu_0 \cos(\theta)}{\epsilon_{\text{eff}} \epsilon_0 A}. \quad (2.4)$$

Only the component orthogonal to the surface of the permanent dipole moment μ_0 affects the work function. This equation includes the tilt angle θ,

2.2. Self-Assembled Monolayers (SAMs)

the elementary charge e, the electric field constant in vacuum ϵ_0 and the area A which is occupied by the adsorbate. The above described depolarization effect, which depends on the coverage density of the SAM and the polarizability of the functional head group of a single molecule, is considered through ϵ_{eff}. The exact charge carrier fluctuations of the depolarization can be calculated using quantum mechanical density functional theory [57]. Summing up the previous effects, the total change in the work function is described as:

$$\Delta\phi = \Delta_{\text{BD}} + \Delta E_{\text{vac}}. \tag{2.5}$$

A very comprehensive study of all surface potential changes during the different stages of SAM formation is presented by Rissner et al. [58]. This investigation clearly shows that the substitution of the functional head group especially has a great influence on the change in the work function and that the change in the work function of the SAM treated metal surface directly affects the injection barrier to an organic semiconductor.

3. Experimental Details

In this chapter the main methods that have been used during this work are described in detail. The next part gives an in-depth insight about the sample preparation, and finally the used devices and the integrated UHV system are presented.

3.1. Methods

In the following section the measurement techniques which were applied during this work are described. The first part covers the basic principles of Kelvin probe (KP), which was mainly used at the beginning of this work for fast characterization of new materials as well as for the establishment of preparation methods. In the next part the photoemission spectroscopy (PES), infrared spectroscopy (IRS), and atomic force microscopy (AFM) principle will be explained in detail. Especially PES and IRS measurements gained a lot of importance during the work progress.

3.1.1. Kelvin Probe

A Kelvin probe is a device which is used to measure the contact potential difference (CPD) between two materials. The method itself was already introduced in the 19th century by Scottish scientist Sir William Thomson (later known as Lord Kelvin) [59]. The original measurement system is depicted in Figure 3.1. At the beginning it was a contact-based method, which in some cases can be destructive for the samples to obtain the CPD. The theory behind the measurement technique is presented in Figure 3.2, which shows the sample and the tip which are in close proximity to each other, but without electrical contact between them. As long as these two materials stay in this

3. Experimental Details

state, their Fermi levels align at energies corresponding to the respective ϕ work functions (Figure 3.2a). If the two conducting materials with different WFs are connected to each other, the resulting flow of charge leads to the alignment of the Fermi levels, which creates a vacuum level difference V_c corresponding to the CPD between these two materials as depicted in Figure 3.2b. Using a variable "backing potential" V_b as a counter potential to V_c, and monitoring the charge at the sample allows to determine the charge-free point where $V_b = -V_c$ as shown in Figure 3.2c. Nowadays KP systems can perform the measurements without contact with the sample, since the same principle is also valid in a capacitive arrangement. Many additional improvements to the measurement technique were added since Lord Kelvin's discovery, but still, the KP remains a relative measurement method (detailed information about modern KP systems can be found in [60]).

The KP measurements during this work were performed on a single point ambient KP system (KP-020 from KP Technology depicted in Figure 3.3) using a 2 mm diameter standard gold coated tip. The resolution of this system is in the range of a few 10 meV. In order to estimate the absolute WF of a measured conductive material, a reference with a very stable WF has to be measured in advance. At the very early phase of this work, the supplied gold reference was used for this purpose. Later, a much more stable and easy-to-clean reference was used, namely highly oriented pyrolytic graphite (HOPG - obtained from Tectra GmbH). The surface can be reproducibly renewed just by pressing a sticky tape to the HOPG piece and peeling it off [61]. The WF of HOPG remains very stable due to the fact that HOPG does not form strong interface dipoles with typical ambient adsorbates like water or hydrocarbons. The WF of used HOPG was measured at the beginning with ultraviolet photoelectron spectroscopy (UPS) and it was 4.47 eV(see Appendix A, Figure A.1).

3.1.2. Photoemission Spectroscopy

Photoemission spectroscopy (PES), also known as photoelectron spectroscopy is one of the most important methods for the investigation of chemical composition and electronic structure of surfaces as well as interfaces between them. The following section is based on the PhD Thesis from Dr. Eric Mankel [62] and [63, 64]. This approach utilizes the external photoelectric effect which

3.1. Methods

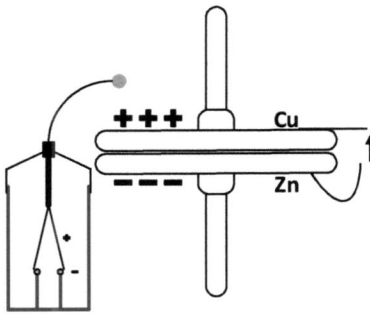

Figure 3.1: Lords Kelvin's original Apparatus [59].

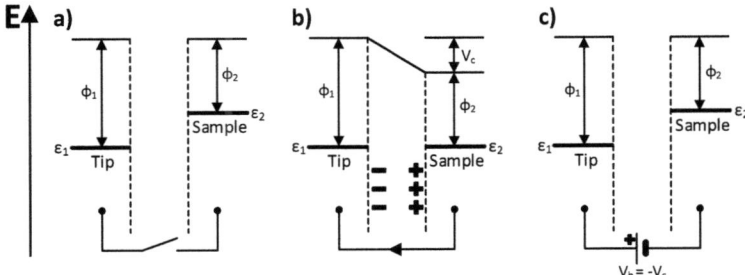

Figure 3.2: The figure shows various electron energy diagrams for sample and tip: a) The electron energy level diagram, where ϕ_1 and ϕ_2 are the work functions of the tip and the sample, respectively. ϵ_1 and ϵ_2 represent their Fermi levels. b) Tip and sample are brought in contact which leads to a net electric current flow until the Fermi level alignment is reached. This has the consequence of a shift of V_c in the vacuum potential. c) Variable "backing potential" V_b is applied to the circuit until $V_b = -V_c$. At this point the vacuum levels align and the CPD between the tip and the sample is found.

3. Experimental Details

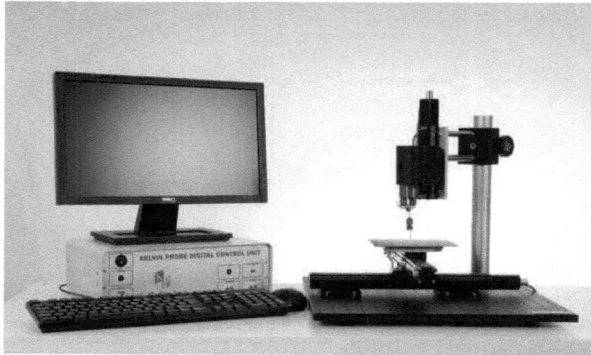

Figure 3.3: KP-020 - Kelvin probe measurement setup [60].

has been discovered and introduced in 1905 by Albert Einstein [65]. This phenomenon is the underlying principle of PES, where illumination of a sample with photons of defined energy (which must be larger than the ionization energy) causes ejection of the electrons into the vacuum. By measuring the kinetic energy of this photoelectrons and comparison to the excitation energy of the light source the binding energies of the electrons can be determined. The working principle can be explained on the basis of the schematic depicted in Figure 3.4. An incident photon releases an electron from the inner shell. The law of conservation of energy allows to calculate the binding energy of the electron:

$$h\upsilon = E_{\text{Bind}} + E_{\text{Kin}} + \phi_{\text{S}} \Rightarrow E_{\text{Bind}} = h\upsilon - E_{\text{Kin}} - \phi_{\text{S}}. \tag{3.1}$$

The binding energy is always referred to the Fermi level as the sample and the spectrometer are electrically connected, which provides also a requirement on the investigated sample; it must be sufficiently conductive. This leads to the conclusion that the electron on the way to the spectrometer must overcome the potential difference between the sample ϕ_{S} and the spectrometer ϕ_{Spec}, thus the binding energy can be written as:

$$E_{\text{Bind}} = h\upsilon - E_{\text{Kin}} - \phi_{\text{S}} + (\phi_{\text{S}} - \phi_{\text{Spec}}) \Rightarrow E_{\text{Bind}} = h\upsilon - E_{\text{Kin}} - \phi_{\text{Spec}}. \tag{3.2}$$

It is therefore independent of the sample work function. By determining the work function of the spectrometer ϕ_{Spec}, which can be easily done by measuring a sample with a known work function (during this work a silver standard was used), the binding energy can by calculated using the equation 3.2.

Depending on the wavelength of the incident light, one can distinguish between X-ray-induced and UV-induced photoemission spectroscopy, which are known as X-ray photoemission spectroscopy (XPS) and ultraviolet photoemission spectroscopy (UPS). Using UPS, a detailed information about valence states as well as the work function of the sample can be obtained. At low kinetic energies, the intensity of the photoelectrons increases strongly as the proportion of inelastic scattered and secondary electrons is increasing continuously. The maximum binding energy of a photoelectron that could be measured corresponds to the light energy $h\nu$. However, these electrons cannot leave the sample, as they need to overcome the work function of the sample ϕ_S first. There are no detectable electrons at this point, only at lower binding energies the intensity suddenly rises. This point is referred to secondary electron (SE) cutoff. These electrons have zero energy at the surface, thus the difference between the light energy $h\nu$ and the SE cutoff $E_{\text{Bind(SE)}}$ is equal to the work function of the sample. It can be written as:

$$E_{\text{Bind}} = h\nu - E_{\text{Kin}} - \phi_\text{S} \qquad E_{\text{Bind(SE)}} \hat{=} E_{\text{Kin}} = 0 \qquad (3.3)$$
$$\Rightarrow \phi_\text{S} = h\nu - E_{\text{Bind(SE)}}.$$

In case of XPS, a detailed information about deeply bound core level states can be obtained, which are specific for the given chemical element and do not contribute to the chemical bonding in a compound. The binding energy of the core level depends also on the chemical surrounding of the atoms, which is referred to as chemical shift.

The photoemission spectroscopy is a very surface sensitive technique. It can be seen as an advantage and disadvantage at once, but can be explained with the inelastic mean free path λ_e of the primary electrons in solids. A general behavior is depicted in Figure 3.5, where the mean free path of the electrons in various materials as a function of kinetic energy is plotted. For small kinetic energies the λ_e is around 25 Å, it reaches its minimum (5 Å) at a kinetic energy of 50 eV and then increases and follows a square root dependence on energy. The information depth which can be obtained with PES reaches approximately 3 to 5 times the mean free path [66, 67]. All electrons from deeper regions are in practice inelastically scattered and contribute only to the background intensity or to satellite emissions. Furthermore, PES can be applied for layer thickness determination if the layer on top is thin enough. A layer with thickness d which is deposited on a given substrate reduces the integral intensity of the substrate emission line I_0. Using this information one can calculate the damped intensity I, which is given by:

$$I = I_0\, e^{-d/\lambda}. \tag{3.4}$$

The PES measurements during this work were performed with the commercially available VERSAPROBE II PES system of PHI. The spectrometer is equipped with a monochromatized Al-K$_\alpha$ X-ray source, an Omicron HIS 13 helium discharge lamp, and a concentric hemispherical analyzer. Detail spectra of the core level lines were recorded with a pass energy of 11.75eV. The spectra and secondary electron edges are referenced in binding energy with respect to the Fermi edge and the core level lines of in situ cleaned Au and Ag samples. For the evaluation of the measured spectra IGOR PRO ver. 4.09 software was used. A more comprehensive description of the PES technique as well as other areas of application can be found in [63]. PES experimental setup used in this work can be found in [68].

3.1.3. Infrared Spectroscopy

One of the most powerful tools for organic thin film investigation is infrared (IR) spectroscopy. IR spectra can be obtained in transmission (Figure 3.6) and reflection (Figure 3.7) mode. In order not to disrupt the scope of this work only a short description of transmission of light in thin-film systems as well as the description of the method itself will be given. The detailed description about the theory behind can be found in [69], [70] and [71].

3.1. Methods

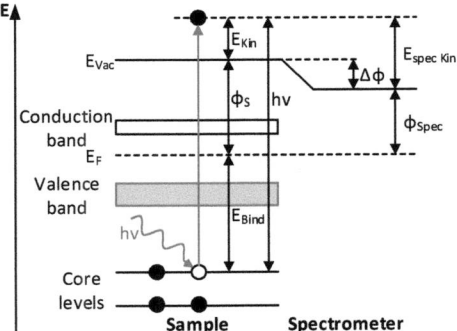

Figure 3.4: The working principle of photoemission spectroscopy. A photon encounters an electron from an inner shell. The photon energy $h\nu$ is completely transferred and distributed into the binding energy E_{Bind} and the kinetic energy E_{Kin} of the electron. As an additional contribution, the work function of the sample ϕ_S must be overcome.

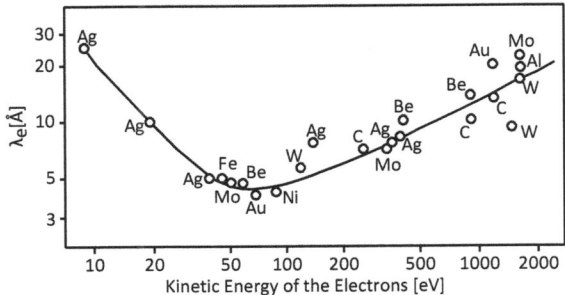

Figure 3.5: Inelastic mean free path of electrons in selected solids. Taken from [64].

3. Experimental Details

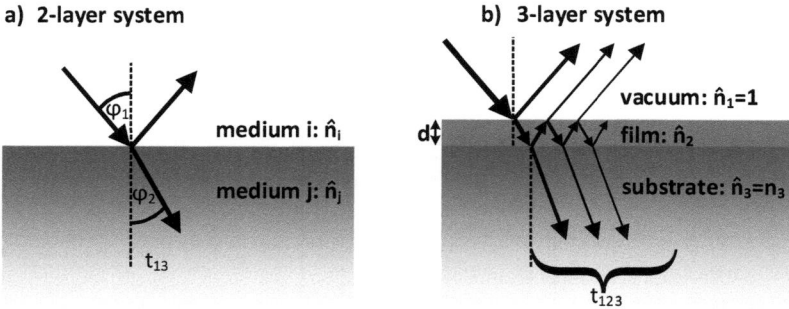

Figure 3.6: Schematic representation of: a) reflection and transmission in two-layer system; b) reflection and transmission in three-layer system.

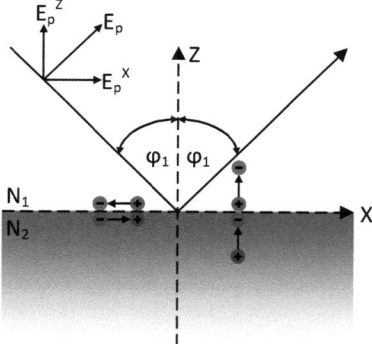

Figure 3.7: Reflection geometry of p-polarized light and schematic representation of the surface selection rule for idealized dipoles. Dipoles parallel to the surface cancel each other, while the perpendicular dipoles are amplified by the electrons in the metal.

3.1.3.1. Transmission of Light in Three-layer Systems

During this work various substrates with different organic thin films were measured using IR spectroscopy, therefore the transmission of a three-layer system in thin film approximation will be derived, based on [69]. In Figure 3.6 such a system is presented. If an electromagnetic wave with an incident angle φ_1 hits the interface between two media with complex refractive indices $\hat{n}_1 = n_1 + i\kappa_1$ and $\hat{n}_2 = n_2 + i\kappa_2$, then one part of the incident radiation is going to be reflected at a given angle φ_1, while the second one enters the medium $\hat{n}_2$ at a given angle φ_2 as shown in the left part of the Figure 3.6. The two angles φ_1 and φ_2 are linked via Snell's law to the refraction indices of the two media:

$$\frac{n_1}{n_2} = \frac{\sin(\varphi_1)}{\sin(\varphi_2)}. \tag{3.5}$$

When a thin film with thickness d is deposited on the substrate (assumption: substrate is downward extended infinitely), then a three-layer system with two interfaces is created. In this case the incident electromagnetic radiation is going to be partly reflected or transmitted on the interface between vacuum and the thin film as well as on the interface among thin film and the substrate, depicted in Figure 3.6b. The reflection and transmission of such a system can be calculated using Fresnel formula, which describe the reflection coefficients as the ratio reflected to incident electromagnetic radiation $r_{ij} = \frac{E_r}{E_e}$ or in case of transmission coefficients, as the ratio of transmitted to incident electromagnetic radiation $t_{ij} = \frac{E_t}{E_e}$ at the interface between two media i and j. The derivation can be carried out from Maxwell's equations taking into account the continuity of the electromagnetic field parallel to the interface and the energy flow perpendicular to the interface. At the same time a distinction between s- and p-polarized[1] light is made. For light propagating in medium i and on the interface with medium j the Fresnel formulas can be written as [69]:

$$t_p = \frac{2\hat{n}_i \cos(\varphi_i)}{\hat{n}_j \cos(\varphi_i) + \hat{n}_i \cos(\varphi_j)}, \tag{3.6}$$

---

[1] S-polarized light - describes light polarized perpendicular to the plane of incidence.
P-polarized light - describes light polarized parallel to the plane of incidence.

3. Experimental Details

$$t_s = \frac{2\hat{n}_i \cos(\varphi_i)}{\hat{n}_i \cos(\varphi_i) + \hat{n}_j \cos(\varphi_j)}, \tag{3.7}$$

$$r_p = \frac{\hat{n}_j \cos(\varphi_i) - \hat{n}_i \cos(\varphi_j)}{\hat{n}_j \cos(\varphi_i) + \hat{n}_i \cos(\varphi_j)}, \tag{3.8}$$

$$r_s = \frac{\hat{n}_i \cos(\varphi_i) - \hat{n}_j \cos(\varphi_j)}{\hat{n}_i \cos(\varphi_i) + \hat{n}_j \cos(\varphi_j)}. \tag{3.9}$$

Taking into account the multiple reflections which can occur in the thin film, the transmission coefficient t_{123} of the three-layer system is given by a geometric series:

$$\begin{aligned} t_{123} &= t_{12}\,e^{i\beta}t_{23} + t_{12}\,e^{i\beta}\,r_{23}\,e^{i\beta}\,r_{21}\,e^{i\beta}\,t_{23} + t_{12}\,e^{i\beta}\,r_{23}\,e^{i\beta}\,r_{21}\,e^{i\beta}\,r_{23}\,e^{i\beta}\,r_{21}t_{23} + \ldots \\ &= t_{12}t_{23}\,e^{i\beta}\left[1 + r_{21}r_{23}\,e^{2i\beta} + (r_{21}r_{23}\,e^{2i\beta})^2 + \ldots\right] \\ &= \frac{t_{12}t_{23}\,e^{i\beta}}{1 - r_{21}r_{23}\,e^{2i\beta}}. \end{aligned} \tag{3.10}$$

Similar applies to the r_{123}:

$$r_{123} = \frac{r_{12} + r_{23}\,e^{2i\beta}}{1 - r_{21}r_{23}\,e^{2i\beta}}. \tag{3.11}$$

This relationship applies to both s- and p- polarized light. By the factor $e^{i\beta}$ the phase difference is taken into account that the light undergoes by a single pass through the deposited layer and can be described as:

$$\beta = \frac{2\pi d}{\lambda}\sqrt{n_2^2 - n_1^2 \sin^2(\varphi_1)} \tag{3.12}$$

With the additional assumptions that medium 1 is vacuum ($\hat{n}_1 = 1$) and that the substrate does not absorb and therefore has a real refractive index $\hat{n}_3 = n_3$ (see Figure 3.6b). For the thin film on the substrate a complex refractive index $\hat{n}_2$ is assumed. Furthermore, the transmittance under normal incidence is considered. Thus determination between s- and p- polarized light can be omitted. The outcome of this is:

$$\varphi_1 = \varphi_2 = 0° \Rightarrow \cos(\varphi_1) = \cos(\varphi_2) = 1 \tag{3.13}$$

and using these assumptions the Fresnel formulas can be simplified to:

$$t_{ij} = \frac{2\hat{n}_i}{\hat{n}_i + \hat{n}_j}, \qquad (3.14)$$

$$r_{ij} = \frac{\hat{n}_j - \hat{n}_i}{\hat{n}_i + \hat{n}_j}. \qquad (3.15)$$

Moreover, the phase difference can be written as:

$$\beta = \frac{2\pi d}{\lambda}\hat{n}_2. \qquad (3.16)$$

The Equation 3.10 is now given by:

$$t_{123} = \frac{2\hat{n}_2}{(\hat{n}_2 + \hat{n}_2 n_3)\cos(\frac{2\pi d}{\lambda}\hat{n}_2) - i(\hat{n}_2^2 + n_3)\cos(\frac{2\pi d}{\lambda}\hat{n}_2)}. \qquad (3.17)$$

As during this work only optically very thin layers ($n < 3$ and $d < 100\,\text{nm}$) are going to be deposited and characterized with IR spectroscopy using light sources with wavelengths λ around µm range, thus another assumption is allowed. With $\cos(\frac{2\pi d}{\lambda}\hat{n}_2) \approx 1$ and $\sin(\frac{2\pi d}{\lambda}\hat{n}_2) \approx \frac{2\pi d}{\lambda}\hat{n}_2$ the Equation 3.17 can be written as:

$$t_{123} \approx \frac{2\hat{n}_2}{(\hat{n}_2 + \hat{n}_2 n_3) - i(\hat{n}_2^2 + n_3)\frac{2\pi d}{\lambda}\hat{n}_2}. \qquad (3.18)$$

The transmittance of three-layer system at normal light incidence, with a transparent substrate and neglecting terms that are quadratic in d/λ can be obtained from:

$$T_{123} = n_3 \left|t_{123}^2\right| \approx \frac{4n_3}{(1+n_3)(1+n_3+\frac{4\pi d}{\lambda}\epsilon_2^{''})}, \qquad (3.19)$$

and without thin film:

$$T_{13} \approx \frac{4n_3}{(1+n_3)}. \qquad (3.20)$$

Using Equations 3.19 and 3.20 the relative transmission of the deposited layer based on the transmission of the pure substrate can be obtained:

3. Experimental Details

$$T_{\text{rel}} = \frac{T_{123}}{T_{13}} \approx \frac{1}{1 + \left(\dfrac{\frac{4\pi d}{\lambda} \epsilon_2''}{1 + n_3}\right)}. \tag{3.21}$$

Taylor series of the first order results in the following approximation:

$$T_{\text{rel}} \approx 1 - \frac{4\pi d \tilde{v}}{1 + n_3} \epsilon_2'' \approx 1 - \frac{4\pi d \tilde{v}}{1 + n_{\text{substrate}}} \epsilon_{\text{film}}'' \tag{3.22}$$

with $\tilde{v} = 1/\lambda$. The relative transmission of the thin film depends in the first approximation on refractive index of the substrate $n_{\text{substrate}}$, imaginary part of the dielectric function ϵ_{film}'' and on thickness of the deposited layer. A detailed derivation of the equations can be found in [69], [72] and [73].

3.1.3.2. FTIR-Spectroscopy

The Fourier transform infrared (FTIR) spectrometer is the further development of IR spectroscopy, where the measurements are not separated by wavelength anymore, but an interferogram from all wavelengths is simultaneously recorded. By applying the Fourier transform the spectrum can be calculated. A short description of this method will be given in this section.

The heart of the FTIR spectrometer is an interferometer, which is an optical assembly to separate light beams (waves), shift them spatially to each other (phase shift) and to overlap them again (interference). FTIR spectroscopy uses the Michelson interferometer which is schematically depicted in Figure 3.8. The incident, broadband IR light splits at the active layer of a beam splitter into two parts with preferably equal intensities. One part is reflected to a fixed mirror and the other one is transmitted to the moving mirror. Both parts are then reflected back to the beam splitter and recombined. Corresponding to the position of the movable mirror, both sub-beams have a path difference $\triangle x$, which leads to interference. The incident light is then reflected to the sample and to the light source. The part which reaches the sample can be either reflected or transmitted to the detector as depicted in Figure 3.8 and its intensity is measured as a function of the mirror position $I_D(\triangle x)$. In order to get the spectral decomposition of the IR radiation, Fourier transformation is applied to the obtained interferogram:

$$I(\tilde{v}) = \frac{1}{2\pi} \int_{-\infty}^{\infty} I_\mathrm{D}(\triangle x) \cos(2\pi\tilde{v}\triangle x) d\triangle x. \tag{3.23}$$

The derivation can be found in [74]. The frequency spectrum $I(\tilde{v})$ is strongly influenced by external parameters like radiation source, detector, beam splitter and all impurities in the beam path. In order to exclude these parameters, a reference spectrum T_REF is measured first. Subsequently, the sample is introduced into the beam path and its spectrum T_Sample is measured. In the relative spectrum the disturbing influence of the external parameters is ideally no longer visible. The relative spectrum is given by:

$$T_\mathrm{Rel} = \frac{T_\mathrm{Sample}}{T_\mathrm{REF}}. \tag{3.24}$$

For the reference measurements, an empty channel measurement or the substrate material measurement are usually used. During this work, organic thin films on silicon as well as on gold coated silicon were measured. Since both of them absorb in the IR range a reference measurement of a non treated clean substrate was always taken. The relative spectra obtained during this thesis are the results of quotient of the spectrum of the untreated and treated substrates.

The IR technique was used for both main topics of this thesis, however a different measurement geometry was applied for each topic. In case of SAMs the infrared reflection-absorption spectroscopy (IRRAS) was used to investigate the properties of the monolayers and for the polymers with thermally activated solubility reduction infrared spectroscopy in transmission mode was applied.

3.1.3.3. Infrared Reflection-Absorption Spectroscopy - IRRAS

In this section a brief description of the IRRAS method will be given. For further details about the technique and further data analysis see [75], [76], [77], [78] and [73].

In this measurement method, the IR light is reflected at the metal surface. The sensitivity of this technique is sufficient to measure monomolecular thin

3. Experimental Details

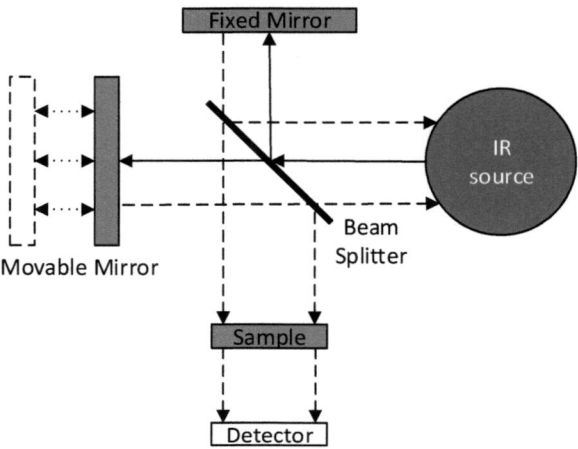

Figure 3.8: Schematic diagram of a Michelson interferometer.

films with a layer thickness of several angstroms. Because of the very high sensitivity of this method not only the chemical analysis can be performed but the main field of application of IRRAS is the orientation determination of the self assembled monolayers. In order to avoid measurement artifacts always the reference spectra of the substrate surface are measured R_{13} and thus the relative spectrum R_{rel} is calculated

$$R_{\text{rel}} = \frac{R_{123}}{R_{13}}. \tag{3.25}$$

Again, in very thin layers where $d/\lambda \leq 0.03$ applies the thin film approximation can be used. Therefore for p-polarized light the following equation is valid, based on [76]:

$$\frac{R_{123,p}}{R_{13,p}} = \frac{8\pi d \cos(\varphi_1)}{\lambda} \text{Im} \left[\frac{\epsilon_2 - \epsilon_3}{\epsilon_1 - \epsilon_3} \left(\frac{1 - \frac{\epsilon_1}{\epsilon_2 \epsilon_3}(\epsilon_2 + \epsilon_3)\sin^2(\varphi_1)}{1 - \frac{1}{\epsilon_3}(\epsilon_1 + \epsilon_3)\sin^2(\varphi_1)} \right) \right]. \tag{3.26}$$

3.1. Methods

Light which is polarized perpendicular to the plane of incidence undergoes a phase shift of approximately 180° on the metal surface. The electric field of the incident and reflected s-polarized wave cancels each other and the phase shift of the light polarized parallel to the incidence plane is 90°, which results in addition of the amplitudes [77]. Therefore, to run IRRAS measurements only p-polarized light is used because the s-polarized components of the electric field cannot contribute to the measurements. Quite the contrary it would just add extra ballast at the detector and thus increase the noise. The p-polarized light also contains a component parallel to the surface E_p^x as shown in Figure 3.7. The incident angle is chosen so that the maximum part of the incident p-polarized light is perpendicular to the surface and the E_p^x disappears [73,77,78]. Assuming that $|\epsilon_3| \gg |\epsilon_2|$ and $\epsilon_1 = 1$ for vacuum and if one consider the incidence angle with the condition that $\cos(\varphi_1) > |\epsilon_3|^{-1}$ the Equation 3.26 can be simplified to:

$$R_{rel} = \frac{R_{123,p}}{R_{13,p}} \approx 1 - \frac{8\pi d \sin^2(\varphi_1)}{\lambda \cos(\varphi_1)} \text{Im}\left(-\frac{1}{\epsilon_2}\right) = 1 - \frac{8\pi d \sin^2(\varphi_1)}{\lambda \cos(\varphi_1)} \frac{\epsilon_2}{|\epsilon_2''|}. \quad (3.27)$$

In comparison to the transmission measurements the reflection measurements may have a sensitivity increased by a factor of $2^{\sin^2(\varphi_1)}/\cos(\varphi_1)$, which is why for reflection measurements usually the largest possible angle of incidence is chosen. Therefore, an incident angle of 80° was used during this work.

All IR measurements presented in this work were performed on commercially available FTIR VERTEX 80v system from BRUKER. A more comprehensive description of the technique, results analysis and the experimental setup can be found in [79] and [74].

3.1.4. Atomic Force Microscopy

Atomic Force Microscopy is a well known and commonly used method to investigate a surface at the atomic level. The method was already introduced in 1986 by G. Binning and coworkers [80]. The working principle is shown in Figure 3.9. The AFM uses as a sensor a thin bar with a fine tip at the front end, called a cantilever. A laser beam is focused at the top of the cantilever and is reflected from there to a position detector. In this way one

3. Experimental Details

can measure the bending of the cantilever. The length of the cantilever is a few hundred micrometers, while the thickness is only a few micrometers, thus, only very small forces are needed to bend the cantilever. This force measurement can now be used to mechanically scan the surface of a sample. The AFM has several operating modes: the simplest is the so-called contact mode (DC mode). Here the bending of the cantilever is kept constant during the whole scan. In order to achieve the constant bending the cantilever is moved up and down with help of a control loop. This movement is recorded with the detector and an image of the surface can be obtained. Here, the forces occurring between the sample and cantilever are in the range of nano-Newtons, which are often too high for the sample as well as for the cantilever. Therefore, nowadays an AFM is normally operated in a tapping mode also known as AC mode. In this mode the cantilever is oscillating continuously at its resonance frequency. Hereby, the laser beam and with that also the detector voltage are oscillating, thus an electrical signal which is proportional to the mechanical oscillation amplitude of the cantilever can be measured. For the excitation with the resonance frequency, only a very small energy input is required. If the oscillating cantilever is brought close to the sample surface, its oscillation is damped already at very low forces between sample and cantilever. The forces occurring between the sample and cantilever are about one to two orders of magnitude smaller than in the DC mode. This leads to less influence of the sample surface and prevents the cantilever from damages. At the same time, this mode of operation has no significant adverse impact on the resolution and is therefore a commonly used mode.

During this work an ambient AFM DUALSCOPE™ DS 95 SERIES from DME[2] was used. The AFM measurements were performed in tapping mode using highly doped silicon cantilevers from NanoWorld (Arrow NCR). These cantilevers have resonance frequencies of about 285 kHz and tip radii of less than 10 nm. A more detailed description about the AFM technique and the experimental setup used during this work can be found under [81]. For the evaluation and processing of the recorded images the supplied DME SCAN TOOL software was used.

---

[2] Danish Micro Engineering A/S, Copenhagen.

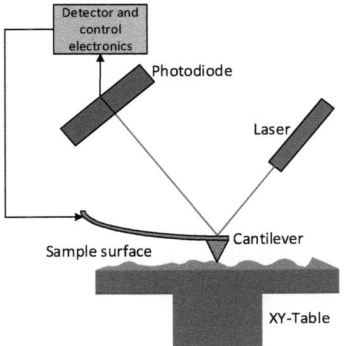

Figure 3.9: Atomic Force Microscopy - schematic representation of the working principle.

3.2. Sample Preparation

In this section a detailed insight into sample preparation will be given. It is split into two main parts. In the first one the sample preparation for treatment with various injection layers (mainly SAMs) is described in detail. An extensive substrate cleaning procedure investigation was carried out in order to obtain very clean and reproducible surfaces. Moreover, the SAM treatment procedure itself will be also shown. In the second part the substrate preparation methods for small molecules as well as for used polymers with thermally activated solubility reduction is described. Finally, the pyrolysis process and its parameters are presented.

3.2.1. Preparation of SAMs

In organic electronics the "cleanliness" plays a very significant role, especially if the interface is going to be characterized. The same applies to SAMs where even the smallest impurities regardless of whether the substrate, solvent or the material itself is contaminated can lead to drastic changes in final physical properties of the characterized monolayer. During this work mainly SAMs with two different anchor groups were used, namely: thiols for metal electrodes

3. Experimental Details

and phosphonates for oxides. In case of phosphonates the sample preparation on indium tin oxide (ITO) was straight forward without complications with impurities or non reproducible results (see 3.2.1.2). On the other hand regarding the thiols and metal substrates many sources describe [33, 40, 82–86] different techniques on how to clean the substrates, how to handle the used solvent or even how to clean the used laboratory glassware. One has to say that thiols are the most studied SAMs in literature. Within the scope of this work, various methods were carried out and most of them lead to very strong changes in achieved results. Moreover, a good part of studies found in literature were carried out on a very well defined gold on mica surface so the knowledge cannot be easily transferred to the evaporated polycrystalline gold surface mostly used during this work. As the biggest impact on the treatment was caused by the substrate's surface itself, a comprehensive study of the cleaning methods was performed in order to get things straight.

3.2.1.1. Gold Substrates Cleaning Procedure Investigation

In order to find a reproducible method to prepare very clean gold surfaces the most promising cleaning procedures found in literature were tested and several XPS and UPS measurements on the treated substrates were carried out. All these measurements were performed in collaboration with TU Darmstadt by Marc Hänsel, a former master student. The full list of these various treatments can be found in Table 3.1.

For each of these methods many different parameters like time of treatment or power of plasma cleaner were tried out, but here only the most representative ones are shown. For PES measurements a silicon wafer with native oxide was used as substrate. All substrates were cleaned with acetone and 2-propanol in an ultrasonic bath as well as treated with oxygen plasma before thermal evaporation. A thin (5 nm) titanium adhesion layer was deposited, followed by deposition of a 150 nm gold layer. The base pressure of the deposition chamber during evaporation was in the 10^{-6} mbar regime, the evaporation rate was approximately 2 Å/s. All results from these measurements can be found in Figure 3.10. For the sake of completeness, all survey spectra of measured samples are presented at the top of the Figure 3.10. The typical gold emission lines can be observed as well as oxygen. Carbon emission lines can be found

3.2. Sample Preparation

Treatment	Description
a: Argon etching (reference sample)	gold reference of the PES system, 4 minutes etching with argon ions at 3 kV acceleration voltage in the UHV chamber
b: Clean room atmosphere	untreated, approx. 30 min in controlled clean room atmosphere
c: Clean room atmosphere (reference sample)	gold reference of the PES system, 30 minutes in controlled clean room atmosphere
d: Oven	heating at 120 °C for 10 minutes in controlled clean room atmosphere
e: Vacuum Oven	heating at 120 °C for 10 minutes in vacuum ~ 10 mbar
f: UV Lamp	10 minutes under UV Lamp in controlled clean room atmosphere
g: Oxygen Plasma	oxygen plasma treatment - various duration
h: Argon Plasma	argon plasma treatment - various duration
i: Argon Plasma & Ethanol	argon plasma treatment - various duration & then 20 minutes immersion in Ethanol
j: Oxygen Plasma & Ethanol	oxygen plasma treatment - various duration & then 20 minutes immersion in Ethanol

Table 3.1: List of gold surface treatment and their description.

3. Experimental Details

in some cases, which gives the first hint about differences in the gold surfaces using different cleaning methods. At the bottom of the same figure, the core level emission lines of O1s, C1s and SE cutoff of Au covered silicon wafers are shown. In the following, the results will be shortly discussed in the order of appearance in the Table 3.1. The sample (special gold foil) with the treatment a is considered as a benchmark for other methods. No contamination is detectable on the surface and the WF corresponds to the literature value of clean gold [87]. In case of the treatment b where the influence of the ambient atmosphere was investigated, after only 30 minutes a significant increase in the carbon and oxygen emission lines can be observed. This leads to a strong change in the WF. In comparison to the reference sample the WF changes by 0.8 eV. No other changes like surface oxidation for example can be seen in the spectra. Using the Au4f emission lines (not shown here) an estimation of the layer thickness of these adsorbates is possible. From the comparison between reference and the untreated sample, an effective surface coverage of 0.48 nm follows. The calculation is based on the assumption of a uniform surface coverage and a free path of electrons in the adsorbate layer of 4 nm. The reference sample was also measured after 30 minutes in ambient atmosphere (treatment c). As expected similar results can be observed. Increase in the carbon and oxygen emission lines as well as a strong change in the WF can be determined. The small deviations may be due to changes in the ambient atmosphere like humidity or temperature during the 30 minutes period. One idea to get rid of the adsorbates was to put the sample into an oven (treatment d) and into a vacuum oven (treatment e). Unfortunately, both methods lead to even more contamination on the surface and to even lower WFs. UV-light radiation is often used in order to clean the surface from various contamination, so this method was also tested (treatment f). As can be seen in the Figure 3.10 the carbon and oxygen emission lines are clearly pronounced and again this is reflected in the work function of the substrate. The treatment g was carried out in a commercially available system - TETRA from DIENER ELECTRONICS. The intensity of the O1s emission line is almost seven times higher than the emission line of the untreated surface which leads to a very small WF of 3.49 eV. A small shift to higher binding energies can be observed in the Au4f emission line of this sample (not shown here), which means the gold

3.2. Sample Preparation

surface is oxidized. Therefore, oxygen plasma treatment has been proven to be unsuitable. The next sample was treated with argon plasma (treatment h), which should not react with the surface. Small peaks in O1s and C1s core levels can be observed and a WF value of 4.8 eV is measured. The method has provided very reproducible results and a WF value which can be repeatedly found in the literature [87]. In [88] it is proposed as a gold pretreatment to first oxidize the sample with oxygen plasma and then to immerse the sample in fresh ethanol for 20 minutes in order to reduce the oxidized gold surface (treatment j). The same procedure was used in treatment i but with argon plasma. In both cases the results were highly unreproducible. In Figure 3.10 only the best results using these treatments are presented (i & j). Sample pretreated with argon shows similar intensity of carbon and oxygen emission lines but lower WF than substrate treated only with argon plasma (treatment h). The highest WF value after the reference was achieved with treatment j, unfortunately this method was proven to be very unstable. Moreover, the level of adsorbates contamination was much higher. In summary, one can see that the increase of adsorbates on the surface scales with a significant decrease of the WF (see Figure 3.11). This phenomenon is known as pillow effect [20,22,89,90] and does explain the effect of lowering the WF. This effect occurs only by physisorption. Once a chemical interaction takes place, the pillow effect is superimposed by the stronger charge transfer in the chemical bound [54,91,92]. Therefore, the cleanliness of the samples plays a crucial role for the WF. In additional measurements the influence of these treatments also on "older"[3] samples were investigated. However, it turns out that the methods can lead to high quality surfaces only on fresh prepared gold samples. Moreover, further optimization of the process led to minimization of impurities and thus the WF value even closer to the in situ reference could be achieved (see Figure 3.12). Unless otherwise specified only the treatment h (argon plasma) was used during the further work. For more information about the cleaning procedure study see [93].

---

[3] Gold samples which were prepared 1-3 days before measurements were taken.

3. Experimental Details

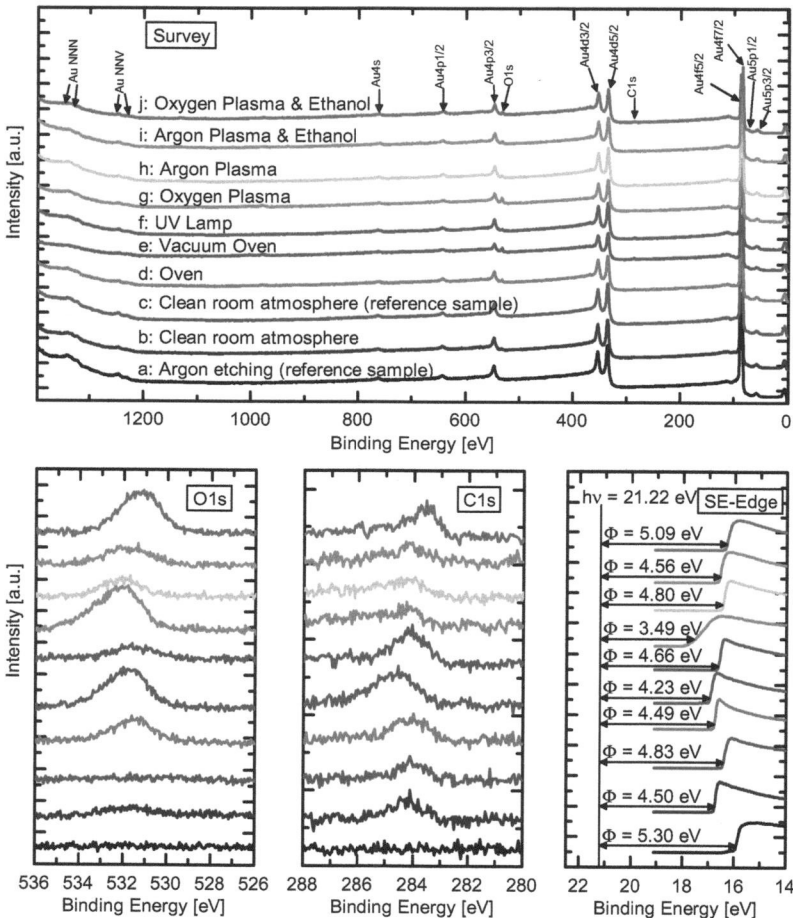

Figure 3.10: Gold XP survey spectrum (top), detail spectra of the O1s (bottom left) and C1s emission lines (bottom center) after the given treatments. The secondary electron (SE) cutoff shows the change in the work function using various cleaning procedures (bottom right).

3.2. Sample Preparation

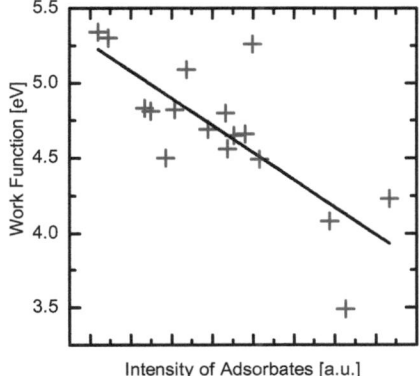

Figure 3.11: Representation of the measured WFs about the sum of the intensities of the C1s and O1s emission lines. The rough trend of the values is shown in black.

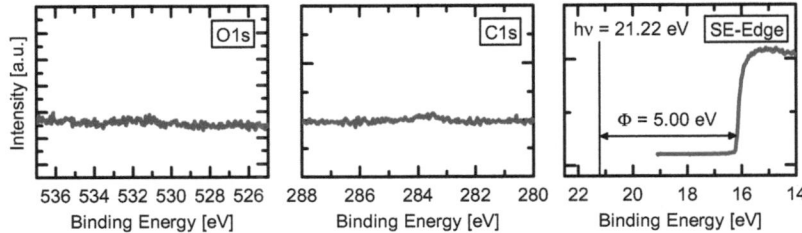

Figure 3.12: Au detail spectra of the O1s (bottom left) and C1s emission lines (bottom center) after optimized argon plasma treatment. The SE cutoff (bottom right) allows to determine the WF of used gold substrate.

3. Experimental Details

3.2.1.2. ITO Substrates

As already mentioned earlier, SAMs with two different anchor groups were used during this work: thiols and phosphonates. In case of phosphonates oxides are needed as a substrate. ITO was the substrate of choice, which is commonly used in various organic devices due to the many interesting properties such as high transparency in the visible, good electrical conductivity and excellent substrate wetting characteristics [66, 94]. Using the ITO cleaning methods found in [66, 95–98] led very quickly to reproducible results. ITO substrates were cleaned with acetone and 2-propanol in an ultrasonic bath for 15 minutes each as well as treated with oxygen plasma for 5 minutes before further functionalization. PES measurements were performed on the clean substrates, the resulting spectra are presented in the Figure 3.13. No substrate contamination can be found in the spectra and the WF value of 4.91 eV corresponds very well to the values found in literature. The reproducibility of this process has proved to be satisfactory. During this work polycrystalline and amorphous ITO was used (see AFM images in Figure 3.14), the differences in the functionality between them will be discussed later (section 5.4.2).

3.2.1.3. Treatment

There are many different methods like physical vapor deposition or electrodeposition on how to treat substrates with a given SAM, however the most commonly used one is the adsorption from solution. Unless otherwise stated, during this work only SAMs prepared by the immersion method were characterized. A simplified picture of this process is presented in Figure 3.15. In case of SAMs with thiol anchor group the clean substrates prepared according to the recipe[4] selected in section 3.2.1.1 are immersed in a dilute SAM solution. After a given time the samples are washed with pure solvent (the same which was used to dissolve the SAM molecules), dried under nitrogen and immediately transferred through ambient conditions to the respective measurement technique. In order to gain the understanding of the self assem-

---

[4] Argon plasma treatment - one needs to mention that prepared substrates are cleaned on both sides with argon plasma in order to remove organic contamination also from the bottom side of the substrate - not needed in case of the previous substrate cleaning procedure investigation.

3.2. Sample Preparation

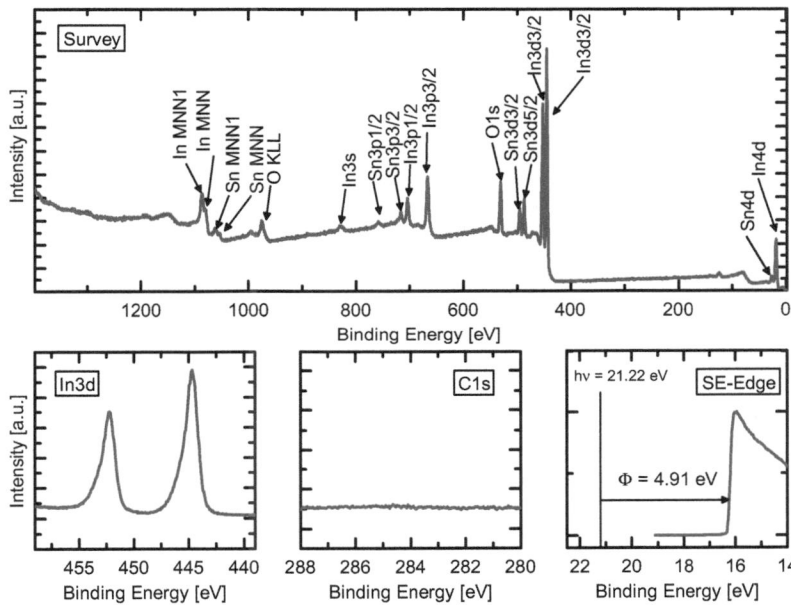

Figure 3.13: ITO XP survey spectrum (top), detail spectra of the In3d (bottom left) and C1s emission lines (bottom center) after oxygen plasma treatment. The WF of ITO can be read out of the SE cutoff (bottom right).

bly process, experiments in various conditions were performed. Experiments were either carried out in ambient atmosphere (in some cases exposed to air for several hours to simulate realistic ambient oxygen conditions) or in a controlled nitrogen atmosphere inside a glovebox. A detailed assignment in which conditions a given experiment was carried out will be always specified during data presentation.

In case of SAMs preparation with phosphonic acids as an anchor group two additional steps are needed. The first one is annealing at 140 °C in order to bond the SAM to the ITO as a monolayer. Finally, any multilayer of a given SAM is removed by sonication in previously used solvent, typically for 30 minutes followed by extensive rinsing with the same solvent and then drying

3. Experimental Details

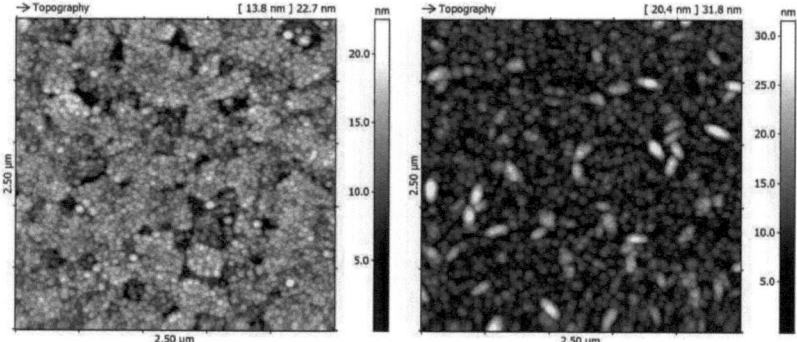

Figure 3.14: AFM images of different ITOs used during this work. On the left polycrystalline ITO can be seen and on the right the amorphous one is shown.

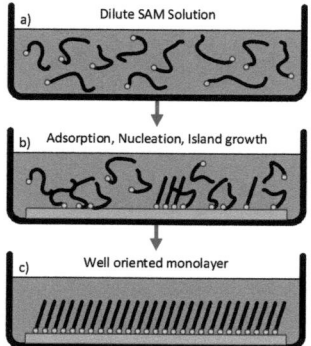

Figure 3.15: A very simplified schematic of monolayer formation on a given substrate. (a) Dilute SAM solution without substrate. (b) After the sample is added, the SAMs begin to physisorb and then chemisorb on the surface. (c) After a sufficiently long immersion time, a homogeneous monolayer should be formed on the substrate surface.

with nitrogen.

3.2.2. Organic Semiconductors with thermally activated solubility reduction

In order to prove the properties of the organic materials, synthesized by Torben Adermann from Organisch-Chemisches Institut (OCI) University of Heidelberg, not only in powder but also as a semiconductor in organic field effect transistors (OFETs), homogeneous thin films need to be prepared, which should be able to bridge the channel between source and drain electrodes. In the present work all new synthesized small molecules or polymers act merely as a precursor for the actual semiconductor material NDI. Therefore, it is necessary that after the thermal conversion of these new NDI derivatives, a homogeneous thin film of the semiconductor material is still obtained. A complete list of tested materials and their formulas as well as initial powder properties like start and end temperature of the pyrolysis process can be found in Chapter 6. Various solvents, solvent mixtures, substrate functionalization and preparation methods were used in order to achieve best results. A detailed assignment on how the layers were prepared will be always specified during data presentation. Unless, otherwise stated all thin films were spin coated from solution.

3.2.2.1. Substrate Preparation

Prior to spincoating, all 25 mm × 25 mm substrates[5] were cleaned for 15 minutes with acetone and 2-propanol in an ultrasonic bath, dried with gaseous nitrogen as well as treated with oxygen plasma for 5 minutes to remove all remaining organic contaminants. All these steps were performed in a clean room and unless otherwise stated in an ambient atmosphere. The prepared samples were then transferred through controlled clean room conditions to the respective measurement technique.

---

[5] Silicon wafer with native oxide and borofloat glass from Schott were used as a substrate.

3. Experimental Details

3.2.2.2. Pyrolysis

The first pyrolysis[6] tests of the prepared thin films were carried out in a special temperature controlled stage (LINKAM LTS420) for Olympus optical microscope, where the sample can be observed and recorded during the pyrolysis process. The stage is optimized for isothermal analysis of large area samples with a thermal stability of better than 0.1 °C. The thin film samples were placed in the heating stage at room temperature to be then heated to the target pyrolysis temperature. Always three different temperature scenarios were used: one with $\sim 5\,\text{K}$ per minute ramp, one with $\sim 20\,\text{K}$ per minute ramp and one where the prepared sample was put directly on the target temperature and hold at the temperature for a given time. Further characterization was only carried out on materials which after pyrolysis still optically looked like closed and homogenous layers. Each sample that survived the pyrolysis process was then tested regarding solubility and measured with different techniques like AFM, ellipsometry (refraction index of 1.8 was used), IR, XPS in order to find out how rough the surface had become and if the cleavable side groups are still on or in the layer. The resulting layer thicknesses of small molecules were measured with profilometer. In some cases scanning electron microscope (SEM) measurements were performed in order to get more information about the layer and its morphology. The NDI derivatives which fulfilled all expectations were then used in an OFET device to investigate its functionality as semiconductor. For this purpose a standard hot plate was used for the pyrolysis process, since the heating stage can accommodate only one sample simultaneously and the comparability between different samples was important. However some temperature adjustments were required due to the change of the hot plate.

3.3. Devices

In this section the devices which were used to prove the functionality of the investigated materials are described. A short working principle description of organic solar cells and organic field effect transistors will be given. Furthermore,

---

[6] Pyrolysis describes a thermo-chemical cleavage of organic compounds, being forced by high temperatures.

3.3. Devices

the used stacks and layouts will be presented and the measuring equipment will be introduced.

3.3.1. Organic Solar Cells

Organic solar cells (OSC), like most organic-based optoelectronic devices, require at least one electrode with sufficiently low WF to have the ability to collect/inject electrons from the electron transporting layer. Low-WF electrodes which meet this requirement are available; however, they are chemically very reactive and oxidize in ambient atmosphere. Within the scope of this work different interfacial layers were used in order to overcome this problem. SAMs with phosphonic acids as an anchor group synthesized within the MORPHEUS project and two polymers PEI and PEIE (chemical structures are presented in Chapter 4) which were introduced in 2012 by Zhou et al. [99] were investigated. Even though the operation mechanism of these polymers is not completely understood, it has not prevented them to establish themselves as reference systems for lowering the WF of various electrodes. During this work a successful attempt was taken to reproduce the results and to gain new insights into the working mechanism. The polymers were used as a reference system because of the simple preparation steps and reproducible effect on WF.

3.3.1.1. Working Principle

A solar cell is a device which can convert the energy of light into electrical energy. It only works if a semiconductor which can absorb light is used, which means that photons excite electrons from the valence band (HOMO level) to the conduction band (LUMO level) [100]. Thereby, excitons[7] are generated. They need to be separated into free charge carriers and conducted out to the selective contacts, otherwise no current flow will be generated. In OSCs the process differ from the one know from inorganic cells [101]. The binding energy of the Frenkel excitons [102, 103] is higher thus much higher electric fields are needed to separate the charge carriers. Hence, in organic solar cells a heterojunction is applied which consists of hole transporting material

---

[7] Electron-hole-pairs bound by the Coulomb interaction.

3. Experimental Details

with relatively high HOMO level and an electron transporting material with a relatively low LUMO level. The principle of this process is depicted in Figure 3.16. A photon is absorbed and forms an exciton, which hopefully comes to the interface and separates into free charge carriers. The separated carriers arrive at the top or bottom contact and are extracted if the energy level alignment between the interface is in good agreement. Since the exciton dissociation occurs at the donator-acceptor interface, the distance between the place where the photon was absorbed and the above mentioned interface must be in the range of the exciton diffusion length. Moreover, the solar cell cannot be arbitrarily thin although the light absorption in organic materials is superb [104]. To meet these boundary conditions the Bulkheterojunction (BHJ) was introduced. In this device architecture the acceptor and donor forms an interpenetrating network which guarantees that the donor-acceptor interface is always in the range of the exciton diffusion length, and thus almost 100% of the excitons are converted into free charge carriers [105]. Certainly, it introduces some other difficulties but they won't be discussed within this work. For further information about the solar cells and their characteristics as well as different device architecture see [28, 106].

3.3.1.2. Organic Solar Cells - Design and Stack Used in this Work

Organic solar cells were always prepared using the same process to ensure comparability. Glass substrates with 140 nm thick ITO layer were used. All substrates were cleaned with acetone and 2-propanol in an ultrasonic bath after the structuring with photo lithography (layout presented in Figure 3.17b). Prior to spincoating of the first layer, all 25 mm ×25 mm substrates were treated with oxygen plasma for 5 minutes to remove all leftover organic contaminants and immediately transferred to the nitrogen glovebox. Three different stacks were always prepared: standard BHJ stack (Figure 3.17a bottom), inverted BHJ stack (Figure 3.17a top) and inverted BHJ stack without interfacial layer. For the standard cells PEDOT:PSS and for the inverted one the interfacial layer was spin coated. In the next step, the active material composed of P3HT and PCBM was spin coated. Metal contacts were prepared by thermal evaporation of 0.6 nm lithium fluoride, 100 nm Al for the standard stack and 20 nm molybdenum oxide, 100 nm Al for the inverted stack

3.3. Devices

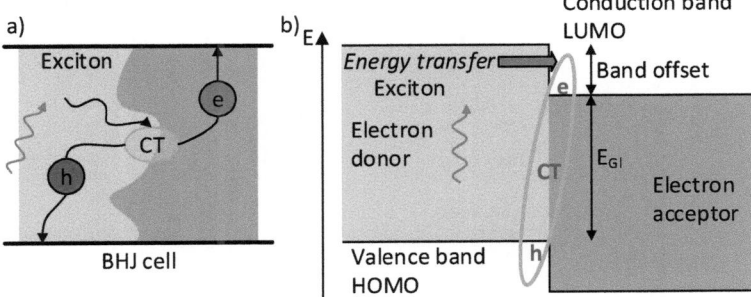

Figure 3.16: Working principle of organic solar cell. a) BHJ cell and its phase separated domain morphology showing photon absorption, exciton diffusion and the so called charge transfer (CT) state which leads to separation into mobile electrons and holes. b) The corresponding energetic scheme of this process. Based on [107].

via shadow mask structuring. The base pressure of the deposition chamber during evaporation was in the 10^{-8} mbar regime. The exact preparation process of BHJ solar cells and their characterization methods can be found in [28, 106, 108].

3.3.2. Organic Field Effect Transistors

OFETs were the next devices used to prove the functionality of investigated interfacial layers. Unless otherwise specified all transistor data shown in this work are prepared in collaboration with the Karlsruhe Institute of Technology (KIT) by PhD student Milan Alt. SAMs with thiol anchor group as well as PEI and PEIE were investigated using OFET devices.

3.3.2.1. Working Principle

The Figure 3.18 shows the origin of the gate-induced charging well known under "field effect" [109]. The positions of the HOMO and LUMO of the organic semiconductor relative to the Fermi levels of the source and drain

3. Experimental Details

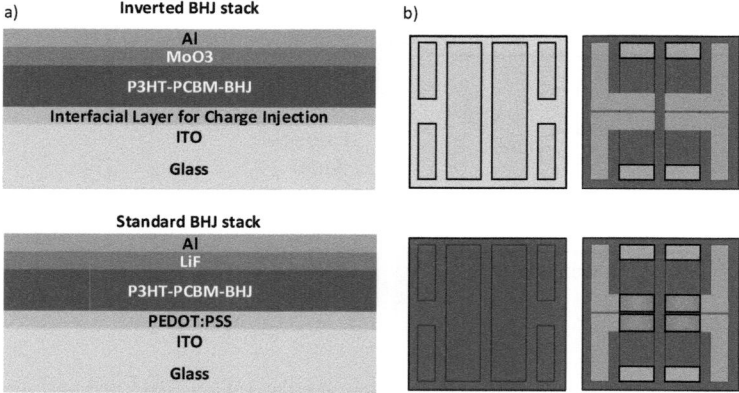

Figure 3.17: Schematic representation of: a) organic solar cell stacks used during this work; (top) inverted stack with interfacial layer on ITO which lowers the WF; (bottom) standard stack. b) used layout - the bright green rectangles represent the active areas (4 mm ×6 mm) of the prepared solar cell.

contacts are drawn in Figure 3.18a. In this scenario the gate bias is $V_G = 0$, so even if a small source drain bias was applied, there would be no conduction because there are no mobile charges in the semiconductor. With a positive gate voltage applied ($V_G > 0$, scenarios b) and d)) a large electric field at the organic dielectric interface is produced, which causes the HOMO and LUMO levels in the semiconductor to shift down with respect to the Fermi levels of the metal contacts. If the gate field is large enough, the LUMO will become resonant with the Fermi levels of the contacts, and electrons can then flow from the contacts into the LUMO. Now there are mobile electrons at the semiconductor insulator interface, which upon application of a drain voltage (Figure 3.18d) result in electric current between the source and drain. This same reasoning applies with negative gate bias (Figure 3.18c, e). Negative gate voltage causes the HOMO and LUMO levels to shift up such that the HOMO becomes resonant with the contact Fermi levels and electrons spill out of the semiconductor and into the contacts, leaving positively charged holes. These holes are now the mobile charges that move in response to an applied drain voltage, Figure 3.18e. Note that in Figure 3.18d, e, the source electrode is always the charge-injecting contact regardless of the sign of the gate voltage. The diagrams are a useful way to visualize the mechanism by which conduction in OFETs is modulated by the gate electrode. However, this description is simplistic and does not take many things into account, like charge traps for example. In addition, the diagrams might lead one to believe that any organic semiconductor can be made to conduct holes or electrons, depending on the sign of the gate voltage. This is not true; in general, a given organic semiconductor can be made more conductive with either a positive or a negative gate voltage, but not both, with only a few recent exceptions. Hence, organic semiconductors are classified according to whether they are hole (p-channel) conductors or electron (n-channel) conductors. For further information about OFETs and their development see [109–112].

3.3.2.2. Organic Field Effect Transistors - Stack and Layout

OFET devices were prepared in staggered bottom contact top gate architecture on glass substrates. Glass substrates were cleaned with acetone and 2-propanol in an ultrasonic bath before thermal evaporation of 60 nm Au source-drain contacts via a shadow mask. All transistors presented feature 20 µm or 50 µm channel length by 1000 µm width. Various semiconductors were used, a detailed assignment about which material was used will be always specified during data presentation. The same applies for dielectric and interfacial layer. However, a general stack is presented in Figure 3.19. One can also say that all semiconductors were spin coated in a nitrogen filled glovebox on top of the substrates and heated afterwards. Gate electrodes were prepared by thermal evaporation of 100 nm Ag via shadow mask structuring. The base pressure of the deposition chamber during evaporation was in the 10^{-6} mbar regime, the evaporation rate was approximately 2 Å/s.

3.3.3. The Clustertool

The clustertool is an integrated UHV System, which allows preparing and characterizing samples using various analytical methods without breaking the vacuum. Figure 3.20 shows a schematic of this integrated system. The

3. Experimental Details

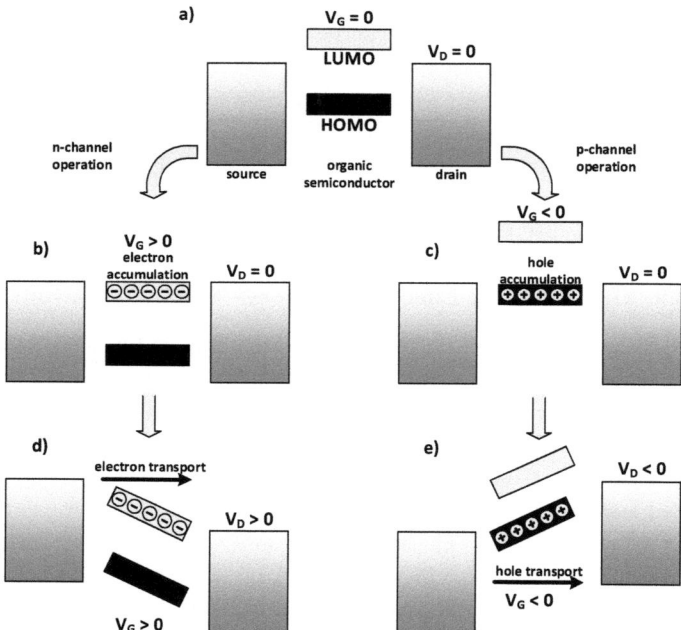

Figure 3.18: Simplified schematic of energy level diagram showing the principle of field effect transistor operation for (left side - b), d)) n-channel operation where electrons are going to be accumulated and (right side - c), e)) for p-channel operation where holes are the majority carriers. Based on [109].

3.3. Devices

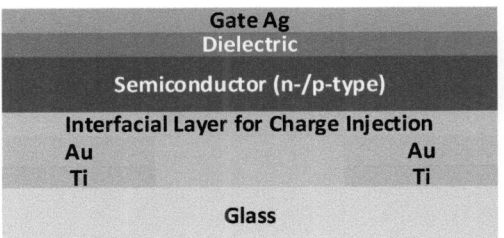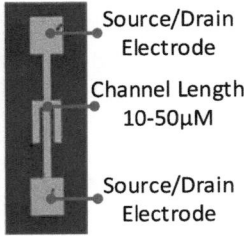

Figure 3.19: Schematic representation of general stack used (left) during this study and a microscope image (right) of a prepared drain/source structure on a glass substrate.

glovebox with nitrogen atmosphere is used for sample preparation purposes, where shadow masks can be changed or a layer can be spin coated from a solution. The glovebox is connected to the clustertool through a load lock. The three transfer chambers (Handler 1-3) provide the possibility to transfer the samples to the target destination without breaking the vacuum. There are three evaporation chambers to choose from two of them are designed as organic evaporation chambers with up to eight evaporation cells each, and the third one is designed for metal evaporation with four sources. The organic films are grown by thermal evaporation using effusion cells which are controlled by adjusting the applied electrical power. In case of an organic evaporation chamber, the temperature of the effusion cells can be also controlled. All the settings are carried out and monitored by MINI8 EUROTHERM controller. The evaporation boats in the metal chamber are directly resistively heated, but unfortunately no temperature control is available in this chamber. The thickness of the evaporated film can be monitored by quartz crystals. In order to ensure very homogeneous film deposition the sample holders are rotated during the whole evaporation process. Such prepared samples can be then transferred through vacuum conditions to several analysis chambers. FTIR spectrometer where IR measurements can be performed, which belongs to the University of Heidelberg, more precisely to the group of Prof. Pucci. The PES system belongs to the group of Prof. Jaegermann from the Technical

3. Experimental Details

University of Darmstadt. SEM combined with focused ion beam (FIB), gas injection system (GIS) as well as scanning probe microscopy (SPM) unit belong to the group of Prof. Kowalsky from Technical University of Braunschweig. Additionally, an independent UHV SPM[8] system is also available.

---

[8] Both SPM systems are capable of performing standard topography measurements (AFM), scanning Kelvin probe microscopy (SKPM) and scanning tunneling microscopy (STM).

3.3. Devices

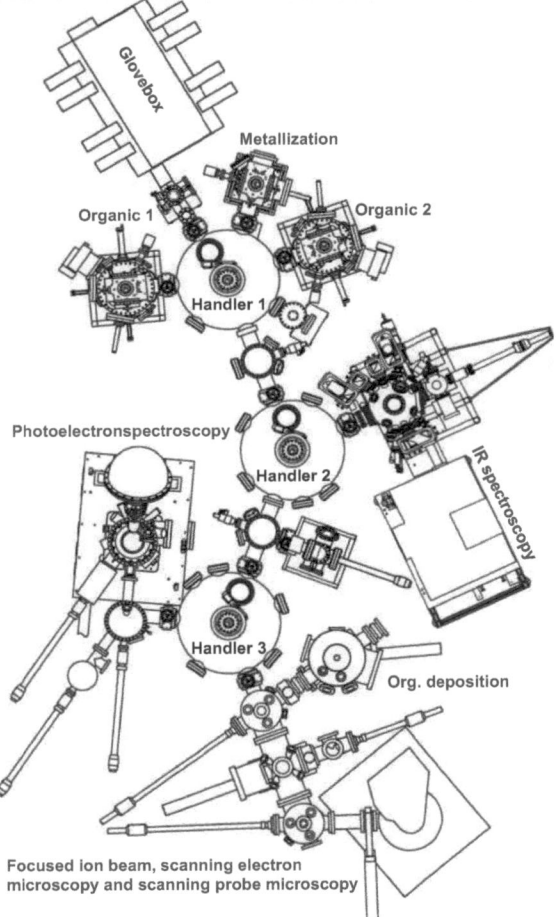

Figure 3.20: Schematic representation of the clustertool at InnovationLab in Heidelberg. All FTIR, PES and SEM measurements were performed on this integrated system. Additionally, for sample preparation glovebox and the three evaporation chambers were used.

4. Polymeric Charge Injection Layers

In the first part of this chapter a short overview about recent works and results of other groups working with polyethylenimine ethoxylated (PEIE) and branched polyethylenimine (PEI) as charge injection layers is given. The next part shows the results achieved in this work. PES and AFM measurements, as well as organic solar cells are introduced. The chapter ends with a comprehensive comparison and discussion.

4.1. State of the art

In organic electronic devices, charge injection at the contacts is crucial for high electrical performance. Most of these devices require at least one electrode with a sufficiently low WF, in order to inject or collect the electrons from the LUMO of the organic semiconductor. Low WF electrodes like alkaline-earth metals are easily available; however, they are chemically very reactive and oxidize in ambient atmosphere. There are different approaches to alter an electrode WF, most importantly transition metal oxides (TMO) [113–115], self-assembled monolayers [50, 116] and the most recent one using polymer charge injection layers (PCILs). PCILs (see Figure 4.1) were already introduced in 2008 by Xiong et al [117] but have not gained the necessary attention. After a very comprehensive study presented by Zhou in 2012 [99] PEI and PEIE became high potential materials even though the underlying mechanisms are still not completely understood. Using these materials allows to reduce the WF of various electrodes in most cases by more than 1 eV. In the work of Zhou et al. various organic electronic devices with PEI/PEIE as interfacial layer were build and in each case they turn the high WF contact into efficient

4. Polymeric Charge Injection Layers

electron-selective electrodes. Zhou et al. claim that the polymers physisorb on the surface and the intrinsic molecular dipole moments associated with the amine groups as well as the charge-transfer character with the conductor surface, make a huge WF reduction possible. Within the next two years several groups reported various electronic devices with inverted structures, however without any additional explanations to the working principle of these interface layers [118–120]. During the collaboration with Sebastian Stolz (PhD student, KIT) an attempt was started to better understand these materials. Stolz et al. could show that the mechanism proposed by Kang et al., which describes a process of electrostatic self-assembly between protonated PEI/PEIE amine groups and oxygen atoms on the electrode surface [121] cannot apply. Unfortunately, the exact working principle could not be explained. Another very recent work already presented during the writing up phase of this study shows that PEI polymer contains N-based impurity molecules with very strong reducing character and that this is responsible for the efficient electron injection [122].

The use of PCILs to tune electrode work functions can be advantageous over SAMs, as the PCIL concept can be applied to a wider range of electrode materials like ITO, Ag, Au or Al without any specific surface chemistry to ensure the functionality. In case of TMOs, such as titanium oxide (TiOx) or zinc oxide (ZnO), which are used as interfacial layer due to their easy solution processability, high annealing temperature (over 200 °C) and additional post-UV treatment (causes often harmful photo-oxidation in the conjugated polymers) is required to enhance the device performance. None of these post treatments are needed in case of PCILs.

The underlying working mechanisms could not be explained within this work, however interesting results are presented. Moreover, the two polymers were established as benchmark materials, thus they are also used to compare the performances with the other interfacial layers used within this study (see Chapter 5).

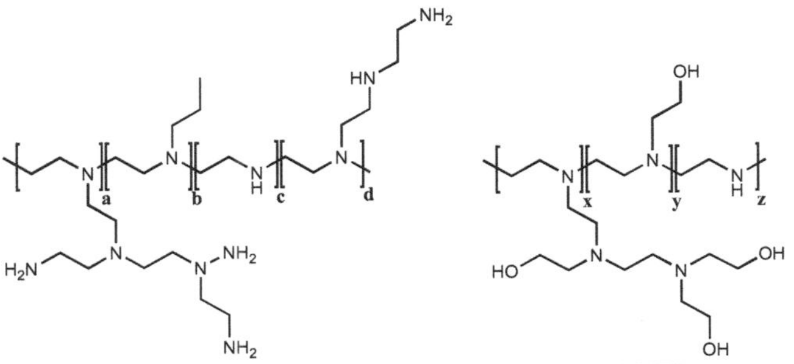

Figure 4.1: Chemical structure of branched polyethylenimine (PEI) and polyethylenimine ethoxylated (PEIE) polymers used in this work and introduced by Xiong in [117].

4.2. PCILs on Various Substrates

As already mentioned above the PCILs are suitable for virtually all used electrode materials. In this work various conducting materials like Ag, Al, Au and ITO were investigated and covered with PEI and PEIE polymers. Silver and aluminum electrodes coated with PEI/PEIE layers are going to be briefly introduced. For more information see the published work of Sebastian Stolz [123]. PEI and PEIE polymers were purchased from Sigma Aldrich and used as received. Al and Ag layers were prepared in the same manner as in case of Au layers described already in Section 3.2. The polymer layers were first dissolved in 2-methoxyethanol (from Sigma Aldrich) with a concentration of 0.4 wt%, spin coated in clean room atmosphere and annealed on a hot plate at 120 °C for 10 minutes. The used spin-coating parameters can be found in [99].

4.2.1. PCILs on Metals - Characterization

Several samples were prepared and measured with the KP system in order to obtain the WF change of the given electrode. The samples were prepared as described in the previous section and immediately measured. Using the KP technique the absolute work function is not measured (for further details

4. Polymeric Charge Injection Layers

see Section 3.1.1) but instead the CPD between the samples. In Figure 4.2 the results of the KP investigation are presented. Each data point is an average from at least four samples, measured at least on two various points within the sample to ensure layer homogeneity. As expected, in each case a significant CPD shift to the lower values is measured. In case of metal electrodes the achieved shift of the WF is not as high as presented in [99]. Further process optimizations were performed, unfortunately without success. Additional studies were performed only on the electrode materials considered as particularly important for this work, like gold and ITO. In Figure 4.3 PES measurements of pure gold and gold samples coated with very thin PEI or PEIE layers are presented. All PES measurement presented in this chapter were performed in collaboration with TU Darmstadt by Eric Mankel. Survey spectra and detailed core levels of these samples were measured in order to verify the chemical composition of the polymers and to examine the pure Au sample. The O1s core level indicates some small contamination of the substrate. However, the C1s emission line of the pure gold samples show that the amount of adsorbates on the surface is negligible, whereas the substrates coated with PEI or PEIE show clearly an increase in the intensity. The samples treated with PEI and PEIE polymers show a symmetric N1s core level, which indicates that all nitrogen atoms have the same oxidation state and that no protonated amine groups are present (similar behavior can be observed for the ITO coated with PEI/PEIE polymer, see next section). This confirms the findings suggested by Stolz et al. The thickness of the resulting layer was too low to be measured by a profilometer, so XPS data were used to estimate the nominal layer thickness (for further theoretical details see Section 3.1.2). The determined layer thicknesses are 1 nm for PEI and 0.6 nm for PEIE, which is in the regime of sub-monolayer thickness. The same samples were characterized with UPS in order to measure their absolute work function (SE-Edge, bottom-right). Even with such a thin polymer layer, a sample coated with PEI has a WF of about 3.77 eV whereas PEIE layer reduces the WF of Au to a value of about 4.1 eV. Several attempts were performed to increase the layer thickness of the PEI/PEIE polymer on Au coated silicon substrate, however without success. The WF shifts measured with UPS however are in the same regime as the ones presented in [99]. The

4.2. PCILs on Various Substrates

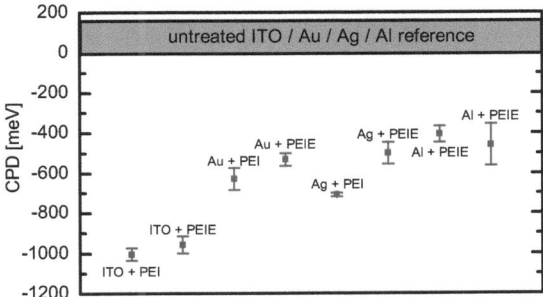

Figure 4.2: KP measurements of PEI and PEIE interfacial layers on different electrode materials are shown. Change in the CPD can be seen, in each case relative to an untreated reference sample. All data points are average values from at least four treated samples, measured at least on two various points in order to ensure layer homogeneity.

discrepancy between KP and UPS measured values, which interestingly is even higher for ITO substrates, and were also confirmed by Sebastian Stolz cannot be explained. It is well known that UPS measurements always yield a slightly lower WF than KP systems but not on this scale. An influence of UHV environment could be excluded.

4.2.2. PCILs on ITO - Characterization

ITO samples coated with PCILs were also investigated in this study. The samples were prepared in the same manner as described in Section 4.2. In [99] an "island" formation of the material on the ITO surface is suggested, which cannot be confirmed by the AFM measurements performed in this study (see Figure 4.4). Almost no difference can be seen between pure ITO substrates and ITO coated with PEI/PEIE polymer. The determined RMS roughness remains also stable at values around 2.3 nm. In the next step, PES measurements were performed on the same sample set. PES measurements were performed by Eric Mankel (TUD). In Figure 4.5 the measured spectra are presented. The survey spectra show all elements within the polymer and ITO. The C1s core level shows a very clean ITO surface as well as the presence of the polymers in

4. Polymeric Charge Injection Layers

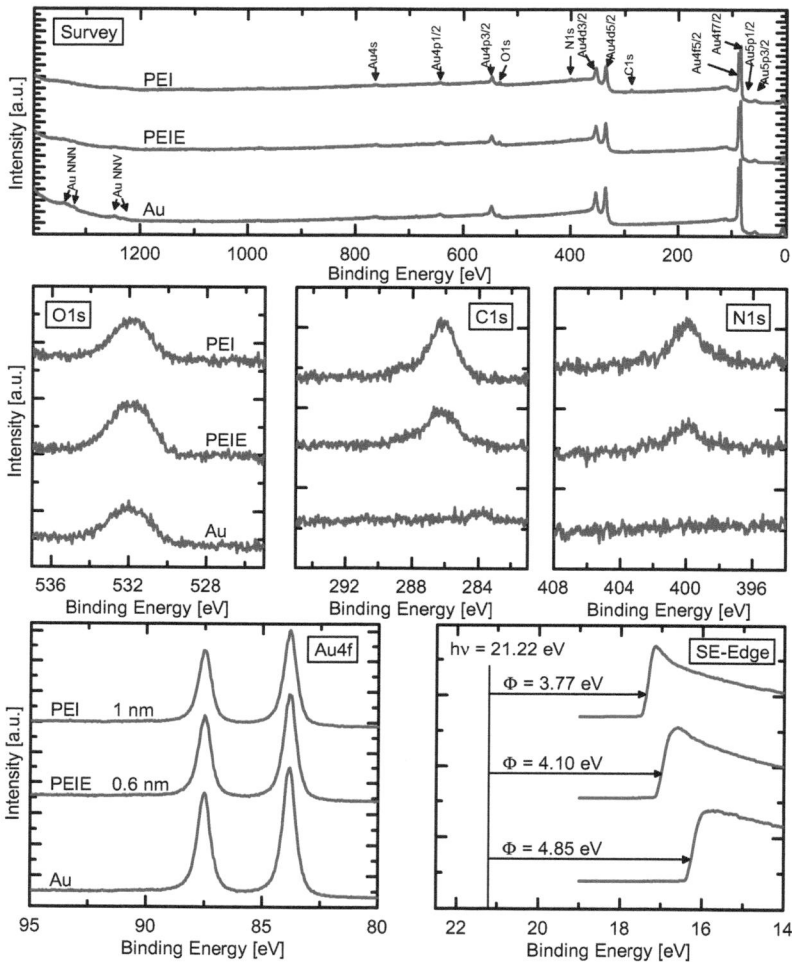

Figure 4.3: PES measurements of pure gold and gold treated with PEI and PEIE polymers. Surveys and the corresponding core levels are shown. Additionally, the SE cutoff shows the change in the work function using PEI and PEIE. The resulting thickness layer is calculated using the Au4f emission lines as described in 3.1.2.

4.2. PCILs on Various Substrates

case of substrates coated with PEI and PEIE. The O1s emission lines indicate the difference between PEI and PEIE polymers. PEIE contains oxygen and the second feature can be clearly seen in this case. The N1 core level shows symmetric peaks, similar to those shown in Section 4.2.1. In the In3d core level of samples coated with PEI and PEIE polymer, especially in the PEIE coated one, a tiny additional feature can be observed. This could indicate a chemical reaction with the ITO surface, like reduction of ITO for example. Moreover, the In3d emission lines were again used to determine the resulting layer thickness. The nominal thickness of the PEI is around 7.2 nm and 3.8 nm in case of PEIE. The layers seem to be thicker than in the case of gold substrate and because both polymers act as insulators such a thick layer could hinder charge transport within a device. For this reason an attempt was taken to prepare thinner layers. Zhou et al. report that the layers can be easily washed off from a given substrate. The two already measured samples were immersed in 2-methoxyethanol (solvent used to dissolve the polymers) and in an ultra sonic bath for one minute. Figure 4.5 shows that this process lead to slightly thinner layers, as depicted by the lower damping of the In3d signal. The estimated thicknesses after "washing" are 6.1 nm for PEI and 2.9 nm for PEIE. All prepared samples were also measured with UPS and the obtained work functions are presented in Figure 4.5 (SE-Edge, bottom-right). The pure ITO has a WF of around 4.92 eV, whereas the samples coated with PEI show a decrease of the WF by about 1.9 eV. The resulting WF of about 3 eV is by 300 meV lower than the result reported in [99] and achieved with a thinner layer. The measured spectra (Au4f and SE cutoff) show that the immersion in 2-methoxyethanol leads indeed to a decrease in the layer thickness, however not to a significant change in WF shift. Also, in this case the comparison between the change in the WF measured by KP system and UPS show even larger discrepancy than in the case of Au substrates coated with PEI and PEIE (see Figure 4.2). With the finding that the layer can be gradually washed off from the treated surface, without significant change in the WF shift, an experiment with different immersion time periods was performed. The results are presented in Figure 4.6. At the beginning of the immersion process even higher WF shift is achieved. This might be explained by the resulting higher homogeneity of the layer. With increasing immersion time,

4. Polymeric Charge Injection Layers

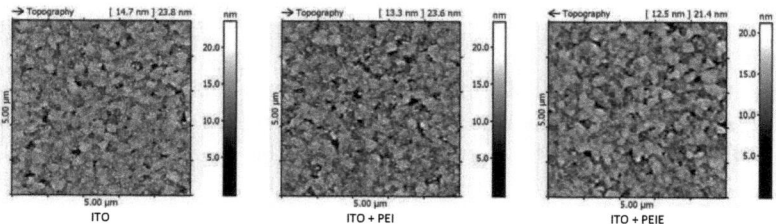

Figure 4.4: AFM images of ITO (left), ITO with spin coated PEI (middle) and ITO with spin coated PEIE (right). No significant change can be seen, no polymer islands on ITO can be observed.

a clear decrease trend in the WF shift is observed, however even after 24 hours immersion in ultra sonic bath, a significant shift of the WF still remains. Longer periods were not investigated, however this result is in contradiction to the one presented in [99] and another hint that the molecules are not only physisorbed on the surface.

4.3. Devices

As a next step a proof of concept was performed on an organic solar cell device. The device layout and used stack for this purpose is described in Section 3.3.1.2. The interfacial layer used within this device was PEI. The resulting devices were than characterized and operated in nitrogen filled glovebox to minimize the influence of oxygen and moisture. In Figure 4.7, the IV characteristics of the measured devices are presented (for more information about the organic solar cells and the measuring system see [108]). The measured curves belong to the hero devices, however a total of four devices for each stack were prepared and in each case the same trend was clearly visible. The key values of standard solar cells are in very good agreement with the usually reached values within the facilities of InnovationLab at this time. In case of the inverted stack, only the solar cell with PEI injection layer provide the same V_{OC} (of course as a negative value) and FF as the standard one. The inverted stack delivers lower V_{OC} and FF values, which indicates an energy loss due to the high WF of non-treated ITO electrode. In Figure 4.8 the corresponding

4.3. Devices

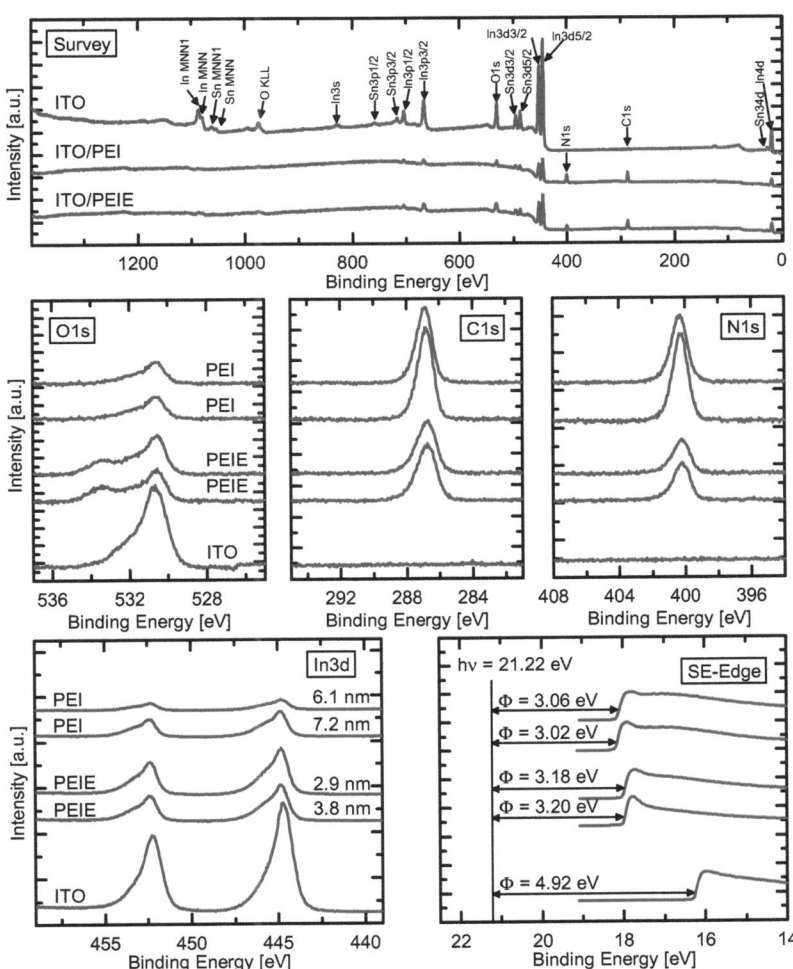

Figure 4.5: PES measurements of pure ITO and ITO treated with PEI and PEIE polymers. Surveys and the important core levels are shown. Additionally, the SE cutoff shows the change in the work function using PEI and PEIE. The resulting thickness layer is calculated using the In3d emission lines as described in 3.1.2.

4. Polymeric Charge Injection Layers

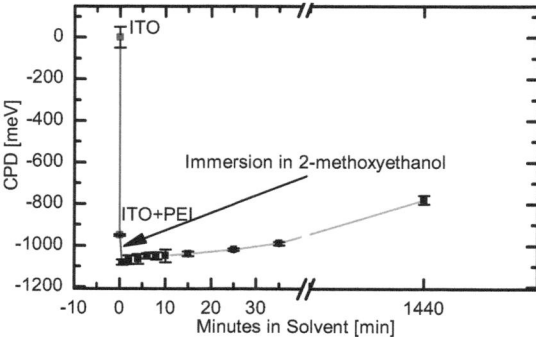

Figure 4.6: Work function shifts of PEI coated ITO surface and the same sample immersed for various time periods in a 2-methoxyethanol ultrasonic bath. Measurements performed with KP system within ambient clean room atmosphere.

energy level diagrams are shown. It must be noted that this representation is very simplified, without any interface interactions, however it is sufficient to explain the effect. In Figure 4.8a the standard stack is presented. The ITO acts as hole-extracting and aluminum as the electron-extracting contact. To improve the injection of charge carriers between ITO and P3HT-PCBM BHJ, a layer of the organic hole conductor material was spin coated, namely poly(3,4-ethylenedioxylenethiophene)-polystyrene sulfonic acid (PEDOT: PSS). Moreover, between the interface of P3HT-PCBM BHJ to aluminum, a 0.6 nm thick doping layer of lithium florid (LiF) was evaporated [124]. In the scheme, the flow of charge carriers is also shown. In the case of inverted solar cells (see Figure 4.8b) the same contact materials and the same P3HT-PCBM BHJ active material is used, however the electrodes are functionalized such that the flow direction of the photo-current is reversed. The WF of ITO is reduced by more than 1 eV using PEI as interfacial layer and thus the ITO electrode can now serve as electron-extracting contact. Between the aluminum and P3HT-PCBM BHJ a 10 nm thick molybdenum oxide (MoO_3) layer is evaporated, for which has been shown that due to its high WF an improvement in the extraction of holes can be observed [125]. An untreated reference was also prepared for comparison (see Figure 4.8c). In this case the ITO electrode

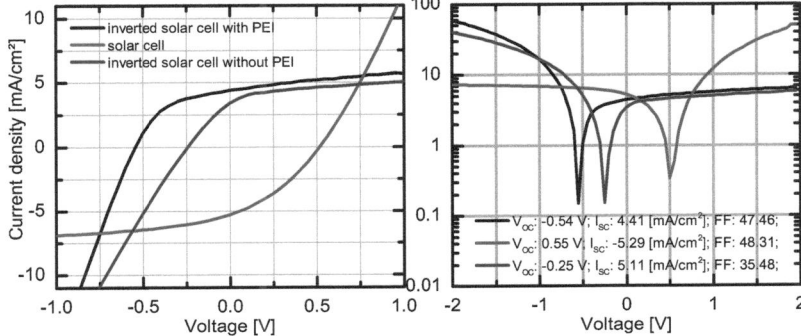

Figure 4.7: IV curves of standard (red) and inverted (blue and black) P3HT/PCBM solar cells under illumination.

was not modified, thus the WF has a value of about 4.9 eV and acts as a large energy barrier for the current flow, more precisely for the electrons.

4.4. Summary

PEI and PEIE polymers were characterized with various analytical methods as well as used as electron injection layers in OSC in combination with ITO electrode. The achieved performance of the OSC devices show that the polymers have a great potential to act as interfacial layer to lower the work function of air stable high WF electrodes even though the polymers themselves act as insulators. One has to mention that no stack optimizations were performed in order to improve the solar cell performance. Various layers of PEI and PEIE were spin coated on different electrode materials and characterized with KP. Additionally, ITO and Au coated with PEI and PEIE were investigated by PES. Both measurement techniques confirmed that these layers reduced the WF of the underlying substrate. The highest achieved WF shift was measured on ITO coated with PEI, with about 1.9 eV. Using XPS measurements, the layer thickness was determined and in case of gold substrates even some sub-monolayers of these polymers lead to significant WF shifts, which can be advantageous for usage in devices. Both polymers

4. Polymeric Charge Injection Layers

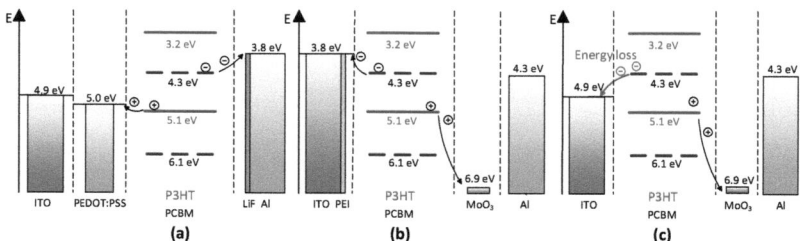

Figure 4.8: Energy level diagrams of P3HT:PCBM solar cells (simplified - interface interactions are not included). The flow direction of the photo-generated charge carriers is marked. (a) Standard configuration where ITO acts as a hole-extracting and Al as an electron-extracting contact. (b) Inverted solar cell where ITO was functionalized with PEI to act as an electron-extracting contact. (c) Reference inverted solar cell, wherein the PEI modification was omitted. The high work function of ITO leads to energy loss. All WF, HOMO and LUMO values, except for ITO and PEI, are taken from [125–128].

provided very homogeneous layers as could be proved by AFM measurements. In this study and in collaboration with Sebastian Stolz (KIT) some of the literature suggested working principles of these polymers were proven false:

1. Kang et al. describe a process of electrostatic self-assembly between protonated PEI/PEIE amine groups and oxygen atoms on the electrode surface. The measurements presented by Stolz et al. as well as the results obtained in this study prove that this cannot apply.

2. Most probably the suggested physisorbtion does not apply either. XPS as well as KP measurements suggest some chemical reaction between the polymer and the underlying substrate.

In collaboration with Milan Alt (KIT), several OFETs were prepared with PEI/PEIE used as injection layer, however these devices are not discussed here. The data will be only used as a comparison to the new SAMs molecules (synthesized by Malte Jesper from OCI within the MORPHEUS project), which are the main topic of the next chapter. A more comprehensive investigation was unfortunately not possible in this work.

5. Self-Assembled Monolayers

This chapter first gives a short overview of recent works and results of other groups working with self-assembled monolayers as injection layers, or as a surface modification layer to improve the wettability for example. The next part presents the results achieved in this work. Several measurement techniques were used in order to characterize the used monolayers. KP, goniometer, PES, and IR results are shown as well as OFET devices with various SAMs, used as injection layer. The chapter ends with a comprehensive comparison and discussion.

5.1. State of the art

Single layers of organic molecules covalently bound to a surface have attracted much attention in the past few years [40, 82, 83, 129, 130]. In 1946 Zisman already reported about the preparation of a monomolecular layer by a self-assembly process [131]. In organic electronics, SAMs are mostly used to tune the work functions of inorganic electrodes used in various devices and thereby controlling the energy barriers for injection or extraction of charge carriers at the interface between electrode and a given active material. In some cases, the used monolayer carries the whole device functionality as the energy alignment on the interface defines the whole charge transport characteristics [132]. Moreover, facile and low temperature processability, compatibility with flexible substrates, reliable film-forming behavior and low costs have meant that SAMs became a widely used method to alter the electrode properties. Several research groups report about changes in the work function of metal electrodes in both directions using polar SAM molecules [50, 133–136]. However, most groups use very long immersion times in order to get well oriented monolayers and are limited to work function

5. Self-Assembled Monolayers

shifts in the range from 4.0 eV → 5.7 eV [55, 116]. Particularly, a stronger reduction of the WF would be beneficial (see Chapter 4) for organic electronics, as well as much faster treatment with immersion times relevant for high-throughput processes like printing. The accepted model for SAM formation differentiates between surface coverage and molecular orientation/ordering. The initial coverage for alkanethiols and phosphonates is completed within seconds [137–140] and starts with a "lying down" phase, followed by a much slower phase, namely the ordering process which can take hours or even days depending on the tail's length and SAM process parameters [83, 141]. It has been shown that phosphonic acid SAMs can be prepared on oxide surfaces with high quality and functionality via spin coating from nonpolar solvents, such as trichloroethylene and chloroform [142–144]. The same is valid for thiol SAM molecules, where numerous studies show short accumulation times on metal surfaces. However most of the groups focus on layer quality like ordering and coverage [138, 140, 145, 146] and not on the influence on WF and their performances in devices. Investigations on SAMs used as injection layers typically describe preparation with long immersion times in order to obtain well saturated SAMs [50, 147, 148]. In this work the focus was set on the relationship between quality and functionality of SAMs prepared from solution and favorably under ambient conditions. The central questions are: What is actually needed to satisfy functionality requirements? What is the influence of the ambient atmosphere and contamination on the self assembly process? Additional degrees of freedom and the impact on the growth, structure and functionality are investigated.

Although this topic seems to be exhausted there is a limitation to the studied substrates, most of the studies are performed on well defined substrates like gold on mica and these are not affordable for use in a device. Moreover, most of them only focus on energy level tailoring, they ignore other parameters crucial for efficient device performances. In case of crystalline semiconductors, the crystalline order at the interface which is important, is in turn dominated by the SAM. Considering printing applications once more, another crucial point is the wetting behavior of the resulting surface. The approach used in this work tries to combine the aforementioned issues whilst still focusing on the energetic effect. New SAM molecules were synthesized by Malte Jesper (OCI, University of Heidelberg) in order to address these questions. All SAMs provided by Malte Jesper were investigated within this study. Various analytical techniques like KP, goniometer, PES, IRRAS were used to investigate the new SAM molecules. Finally the monolayers were used and characterized within OFET devices.

5.2. Preliminary Examination

In order to establish the SAM treatment process several commercially available SAM molecules were purchased by Sigma Aldrich and investigated with KP and goniometer setup at the beginning of this work. The SAMs were prepared on thermally evaporated polycrystalline gold films according to the recipe described in Section 3.2. One has to mention that no argon plasma treatment was performed as at this stage of this work no such treatment was available at InnovationLab. All samples were immersed for 24 hours in precursor molecule dissolved in ethanol at concentrations of 0.1 mM, 1 mM, 3 mM and 10 mM, respectively. The best performances which in this case is defined as the highest WF shift were achieved with the 1 mM concentration. The results are presented in Figure 5.1. The chemical structures of used thiol SAMs are depicted in Figure A.2. The results show that the WF of gold and most probably also of other metal electrodes can be easily shifted by $\Delta\phi \approx -600\,\text{mV}$ as well as to the higher values by $\Delta\phi \approx 1000\,\text{mV}$. The KP Technology supplied gold sample was used as a reference for these measurements (WF value of the supplied gold is around $\phi \approx 4.7\,\text{eV}$, which is a well known value for an Au surface exposed to ambient atmosphere [149]). As expected, the highest shift to the lower values was achieved with the hexadecanethiol. The alkyl groups $(CH_3(CH_2)n)$ have a slight electron-donating effect (also known as positive inductive effect "$+I$") due to the moderate electronegativity of the hydrogen atoms. This leads to polar bond dipoles caused by displaced charge centers. The vector sum of all bond dipoles provides an effective dipole moment. In case of alkylthiols the vector shows along their axis in the direction of the thiols anchor group. Each carbon element increases the length $\vec{l}$ of the dipole moment $\vec{\mu_0}$ and thus the amount referred to $\vec{\mu_0} = q \cdot \vec{l}$. The scaling factor is the acting charge q. The perfluorodecanethiol shows the highest change

5. Self-Assembled Monolayers

in the WF and by replacing the hydrogen with fluorine atoms the direction of the dipole moment rotates and points away from the surface. Fluorine is the element with the highest electronegativity on the Pauling scale, therefore the trifluoromethyl group has a strong negative inductive effect "$-I$" and their environment electrons are attracted with high affinity. This leads to the inverse impact on the work function of the treated substrate. The remaining molecules are not further explained here, for more information see bachelor thesis of Sebastian Hietzschold (University of Heidelberg) [150].

Contact angle measurements of water on monolayers prepared by the immersion in solution show that all used SAM molecules make the gold surface more hydrophobic, which means that the gold samples have lower surface energy (see Figure 5.2). All alkanethiols with aromatic rings as well as the one with two thiol groups lead to slightly higher surface energies. In case of SAMs with various alkyl chain lengths, the hydrophobicity increases with longer chains and reaches contact angle of water around $112° \pm 0.8°$ for hexadecanethiol as already shown in [151]. The fluorine as functional head group amplifies the hydrophobic effect and reaches a contact angle of $115° \pm 0.5°$ and also provides the lowest surface energy. In this work, several devices like simple diodes and OFET devices were prepared and investigated using the various commercially available SAM molecules as injection layers. However, in order not to disrupt the scope of this work these are not going to be discussed here.

In collaboration with Milan Alt (KIT) additional KP measurements on gold substrates treated with various SAM molecules and with different immersion times were performed, in order to obtain the shortest treatment time needed for OFET devices. The results are presented in Figure 5.3. It stands out that the standard long immersion times (24 hours) not necessarily provide the best results (greatest WF shift), without distinction between increasing or decreasing the WF of a given substrate, so functional head group independent. The same applies to OFET devices prepared by Milan Alt, which show a decrease in performances with longer drain and source electrode treatment. In order to investigate this behavior in detail 1H,1H,2H,2H Perfluorodecanethiol was chosen as a model system. This SAM molecule increases clearly the WF and thus the induced change cannot be confused with simple adsorbates contamination, which are normally causing a WF decrease of a noble metal.

5.2. Preliminary Examination

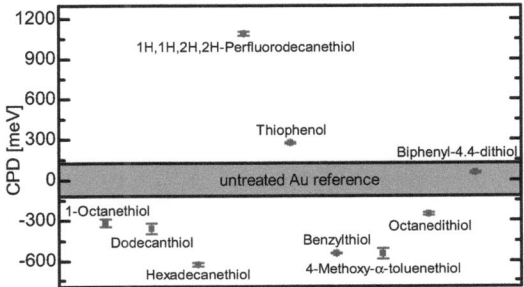

Figure 5.1: The change in the WF of the substrates treated with various SAM molecules relative to an untreated reference sample (Au) measured by KP. Each point represents the average value obtained from at least 4 samples.

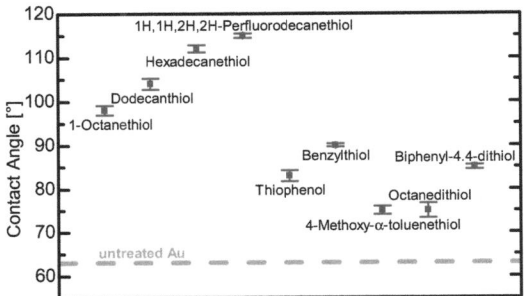

Figure 5.2: Contact angle measurements of bare gold and gold treated with various SAM molecules. Bare gold substrate is the most hydrophilic one, showing a static contact angle of $63° \pm 1.1°$ with water which is in good agreement with the values in literature [152]. Each point represents the average value obtained from at least 4 samples.

5. Self-Assembled Monolayers

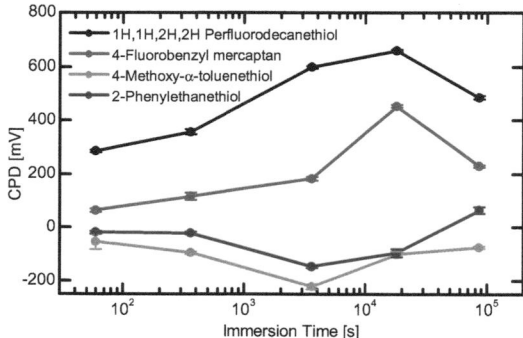

Figure 5.3: KP measurements of gold treated substrates with various SAM molecules using different immersion times.

5.3. Perfluorodecanethiol (PFDT) as Model System

The 1H,1H,2H,2H Perfluorodecanethiol (PFDT) immobilized on Au surfaces was chosen as model system, as this combination has been studied extensively and is known to increase the work function of Au by $\sim 0.5\,\text{eV}$ [153]. In order to investigate the formation process of PFDT SAMs the molecular exposure (defined as product of immersion time of the used substrate and the used SAM concentration) and the processing environment was varied. The treated substrates were characterized by PES and IRRAS, and implemented as charge injection layers in OFET devices. The chemical structure of PFDT is shown in Figure 5.4.

5.3.1. PFDT - PES characterization

All characterized samples were prepared according to the recipe described in Section 3.2.1. One has to mention that for the first PES characterization the preparation was carried out in a nitrogen filled glovebox. The treated substrates were then immediately transferred to the respective measurement technique through ambient clean room atmosphere. As a first step the XPS measurements were performed in order to prove the monolayer presence on

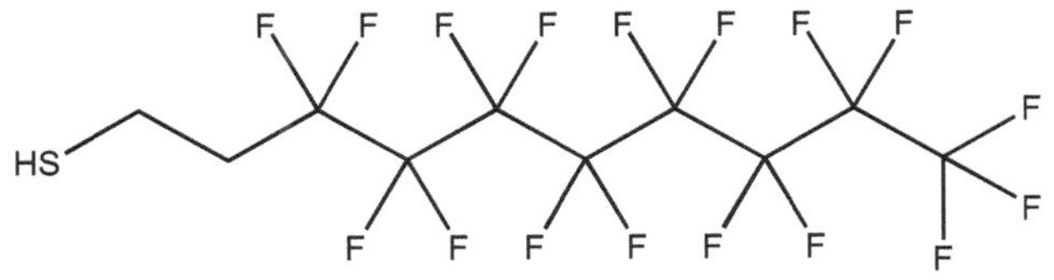

5.3. Perfluorodecanethiol (PFDT) as Model System

1H,1H,2H,2H-Perfluorodecanethiol

Figure 5.4: Chemical structure of 1H,1H,2H,2H Perfluorodecanethiol used as model system.

the surface as well as to assign the core level peaks to the PFDT molecule. In Figure 5.5 the C1s (left) and F1s (right) core level spectra as well as the corresponding fits are presented. The fit and terminology is similar to that proposed in [154]. Black dots correspond to the measured spectra, however with removed background. The red line shows the resulting fit. At the bottom of each graph an error curve is plotted, which describes the deviation between the fit and the measurement (the error curve is 5 times magnified). The C1s emission line shows three different peaks which can be assigned to the three different carbon species within the PFDT molecule: carbon bound to three fluorine atoms (orange), carbon bound to two fluorine atoms (blue) and carbon bound to two hydrogen atoms (green). The one carbon which is bound to three fluorine atoms has the highest oxidation state ($+III$) within this molecule and as expected the singlet corresponding to this carbon atom is chemically shifted to higher binding energies. The same applies to the carbon atoms bound to two fluorine atoms, which have a formal oxidation state of $+II$. The measured binding energy of 284.5 eV which corresponds to the not oxidized or reduced carbon atoms is in good agreement with values found in literature [155]. The intensity ratios between the three carbon species match very well to the stoichiometry of the PFDT molecule. The same applies to the F1s emission line whose measured intensity corresponds very well to the number of fluorine atoms within the PFDT SAM. In the energy range of the S2p orbital (around 162 eV) hardly a doublet structure can be seen (not shown

5. Self-Assembled Monolayers

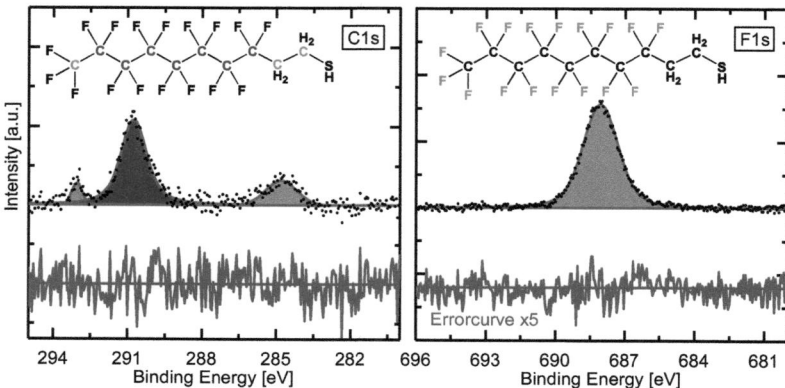

Figure 5.5: Assignment of C1s and F1s emission peaks to the molecular structure of PFDT.

here). Due to the poor signal-to-noise ratio, no quantitative statements are derived from the S2p emission line.

5.3.2. Exposure Dependent Monolayer Formation and Work Function Shift

The next logical step was to prepare several samples with increasing exposure to PFDT solution. Used time steps are presented at the bottom of Figure 5.6. A series of samples immersed in a very dilute PFDT solution (0.004 mM concentration) for $5 \rightarrow 5^7$ seconds was prepared and investigated by PES and IRRAS techniques. In order to cover higher exposures samples with 5^7 seconds immersion time, 0.02 mM and 0.1 mM concentrations were prepared and investigated. All these samples were prepared in inert glovebox atmosphere and immersed in dilute PFDT solution using anhydrous and oxygen-free ethanol solvent as the effect of decrease in functionality described in previous section is most probably caused due to exposure to ambient atmosphere. In the survey spectra (see top of the Figure 5.6) of each sample an increase in PFDT features with increasing immersion time is clearly seen. The F1s detail spectra (center-left) show the increasing coverage of PFDT monolayer until

5.3. Perfluorodecanethiol (PFDT) as Model System

the monolayer formation is completed (after 5^6 s at 0.004 mM concentration). The Au4f emission lines presented in Figure 5.7 were used to estimate the nominal layer thickness (for further theoretical details see Section 3.1.2). The determined layer thickness of the completed monolayer is around 1.1 nm which corresponds very well to the theoretical PFDT length of 1.2 nm under the assumption that the molecule is not "staying" perpendicular to the surface. The C1s detail spectra (see Figure 5.6) show at very low exposures (5 s and 5^2 s) some carbon species which cannot be assigned to the PFDT molecule but most probably to some adsorbates from the solvent. Fortunately, in the next time steps the initial carbon lines are gradually replaced by three different carbon species, which correspond to the chemical composition of the PFDT molecule (see Figure 5.5). A similar effect is observed in the SE cutoff spectra, where the WF is decreasing for the very low exposures. This phenomenon is known for noble metals as pillow or push back effect [20,22,89,90]. First at 5^3 s immersion time when the C1s and F1s emission lines start to show an increase in the intensity the WF for this sample equates to the WF of the Au reference from the beginning and saturates at approximately 5.45 eV for long immersion times. Additionally, in Figure 5.7 the O1s detail spectra, which were measured to monitor the possible oxygen contamination are shown. However, no oxygen was detected and no decrease in the WF shift was obtained, even after almost 24 hours immersion time.

The same set of samples prepared concurrently in the same environment were characterized with IRRAS. All IR measurements were performed and evaluated by Sabina Hillebrandt (University of Heidelberg). The measured spectra presented in Figure 5.8 show strong changes in both relative and absolute intensities of the vibrational modes with increasing immersion time. The samples with shortest immersion times (5 s and 5^2 s) show peak features which cannot be assigned to vibrations of the PFDT molecule. This confirms the observations from PES measurements that at very low exposure times the gold surface is covered with some adsorbates. As described in Section 3.1.3.3 on metal surfaces only vibrational modes with a transition dipole moment component parallel to the surface normal can be excited with IRRAS. For low exposures ($0.5 \, \text{s} \cdot \text{mM} \triangleq 5^3$ s immersion time at 0.004 mM concentration) first characteristic PFDT bands are showing up. At the beginning the orthogonal

5. Self-Assembled Monolayers

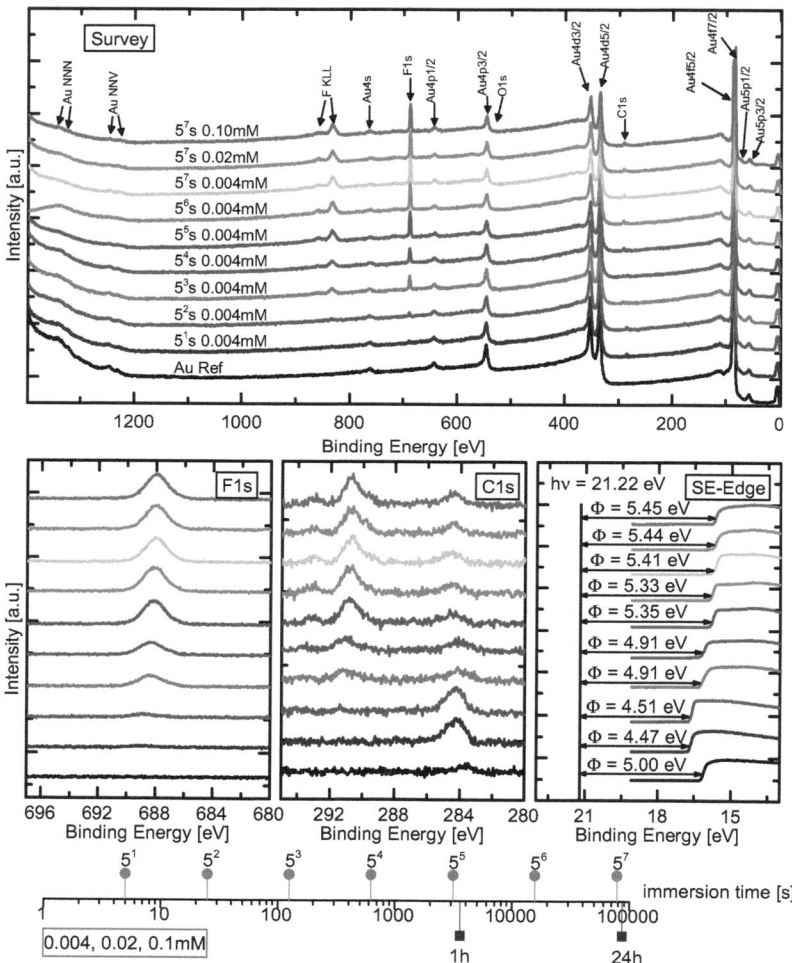

Figure 5.6: XPS and UPS spectra - XP survey spectra of untreated and treated gold samples (top). F1s and C1s emission and SE cutoff with and without SAM treatment (middle), various immersion times and different concentrations (0.004mM - 0.1mM, bottom).

5.3. Perfluorodecanethiol (PFDT) as Model System

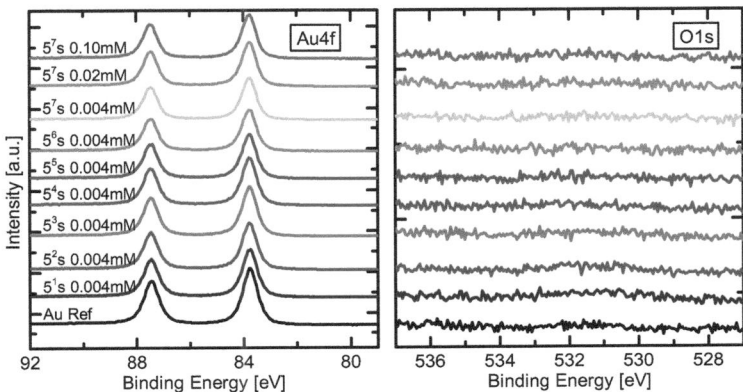

Figure 5.7: Corresponding Au4f and O1s core level spectra of gold samples before and after PFDT treatment for different immersion times.

bands show higher intensities than the parallel ones. This ratio changes for long immersion times ($62.5\,\text{s} \cdot \text{mM} \mathrel{\hat=} 5^6\,\text{s}$ immersion time at $0.004\,\text{mM}$ concentration), where the parallel bands show the higher intensities, which means that the molecular axis of PFDT SAM is approaching the surface normal as expected in the SAM formation process [33, 141]. The whole picture obtained from IRRAS measurements correlates with the corresponding PES measurements described above. Using the relative reflectance R between orthogonal and parallel bands and comparing them with bulk measurements of PFDT a tilt angle θ was derived for the different immersion times. The used relation can be written as:

$$\frac{\log(R_{\text{SAM,p}})}{\log(R_{\text{SAM,o}})} \times \left(\frac{\log(R_{\text{bulk,p}})}{\log(R_{\text{bulk,o}})}\right)^{-1} \times \frac{v_{\text{o}}}{v_{\text{p}}} = \frac{1}{\tan^2(\theta)}. \tag{5.1}$$

The v refers to the wavenumber of the orthogonal (v_{o}) and parallel (v_{p}) absorption bands. For more details see [75, 156]. The determined tilt angle is around $10° \pm 3°$ which is slightly different from reported literature values ($12° \rightarrow 16°$) [157–159]. The deviation depends on various factors, but most probably on the cleanliness of the given substrate and obviously the purity of the used solvent as well as the SAM material itself which plays a significant

5. Self-Assembled Monolayers

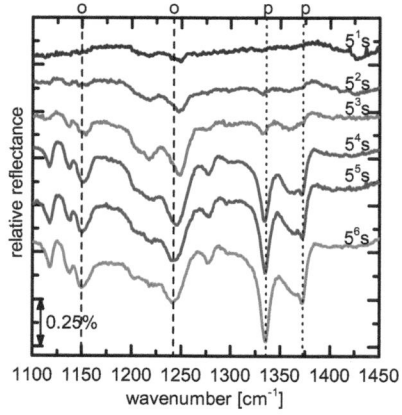

Figure 5.8: IRRA spectra of PFDT monolayers for different immersion times of the gold substrate in a 0.004 mM solution. The dotted and dashed lines mark vibrational modes with the transition dipole moment parallel respectively orthogonal to the molecular axis.

role.

In further experiments similar sets of samples were prepared, however with higher concentration (0.1 mM) and in ambient atmosphere. Only the meaningful O1s and SE cutoff spectra are presented in Figure 5.9. With such a high concentration and clean gold substrates, after 5 seconds immersion time only the expected shift in the WF is achieved and the measured F1s and C1s emission lines show the same intensities as the samples from previous experiment with completed SAM formation. The same applies for the IRRAS measurements. However, more interesting is the fact that the measured WF shift remains almost unchanged as long as no oxygen contamination can be detected on the surface. Clearly, there is a correlation between the achieved WF shift and the oxygen contamination in the PFDT layer. Already after almost five hours, where the oxygen emission line is barely visible, the WF shift is strongly decreased. Moreover, the oxygen presence is not only canceling out the SAM functionality, but it causes even a shift of the WF in the opposite direction (5^7 s immersed sample has a WF of 4.06 eV). This is consistent with the SAM behavior seen in OFET devices (measured by Milan Alt) from exaggerated immersion time in ambient atmosphere. Apart from oxygen contamination no additional contaminants were found on the samples so the loss in functionality obviously corresponds to the oxygen contamination.

5.3. Perfluorodecanethiol (PFDT) as Model System

Additionally, some extreme cases were investigated, where each two samples were prepared in exactly the same manner but in various environments. The first pair was prepared in nitrogen filled glovebox and the second one in ambient atmosphere. One has to mention that exactly the same solution was used, prepared in inert glovebox atmosphere. The first samples were measured after one hour and the remaining two first after seven days. The resulting spectra are presented in Figure 5.10. In case of samples prepared in the glovebox no change in the C1s emission line structure can be observed (left). The peak structure can still be assigned to the three different carbon species within the PFDT molecule. When prepared under ambient conditions (right) the C1s peak structure shows a change between one hour and seven days immersion times. Therefore, the loss in SAM functionality is most probably caused by a change in the molecular dipole induced by oxidative degradation (the additional structure appears on higher binding energies), and not by desorption of the molecules from the metal surface. Oxidative degradation of partially or fully fluorinated hydrocarbons is well known and already studied in the literature, but not in context of SAMs [160, 161]. However, as shown in Section 5.2 Figure 5.3 this issue is not necessarily limited to fluorocarbon containing SAMs but most probably also different SAM molecules. On the other hand, if the whole preparation process is carried out on clean substrates using oxygen-free materials and with high exposure (short immersion times but high SAM concentration) a satisfactory result can be achieved even under ambient conditions. Once the substrate with the SAM is removed from the ethanol solution and transferred in an oxygen-free atmosphere or encapsulated no further decrease of the dipole of the molecule could be observed.

A similar experiment was performed by Milan Alt (KIT) on OFET devices. The devices were always prepared at the same time and using the same solutions as for the samples prepared for PES and IRRAS measurements. The OFETs were prepared according to the recipe and layout described in Section 3.3.2.2. PIF8-TAA (for chemical structure see Figure A.3) [162] was used as semiconductor material and 300 nm of parylene-C served as dielectric. OFETs treated with PFDT SAM are compared to devices with untreated drain and source electrode. Due to improved hole injection the devices with treated electrodes should show lower threshold voltages (V_{th}). V_{th} were evaluated

5. Self-Assembled Monolayers

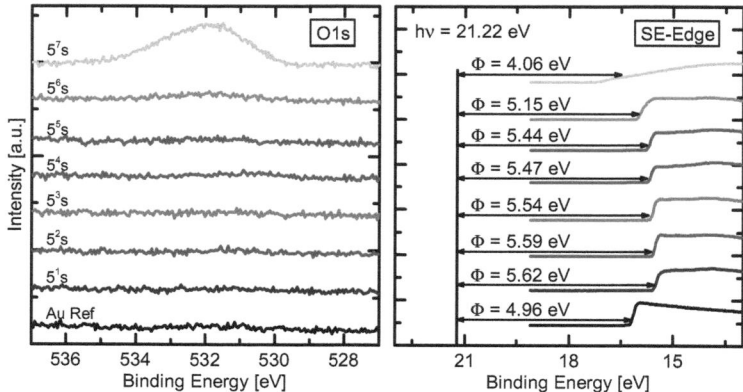

Figure 5.9: O1s emission lines and SE cutoff for different immersion times of the gold substrate. Samples prepared under ambient condition. Already after 5 seconds immersion time only the expected shift in the WF is achieved. The measured WF shift remains almost unchanged as long as no oxygen contamination can be detected on the surface.

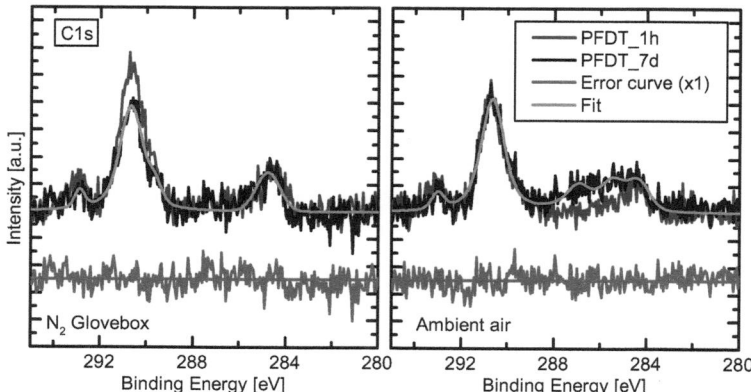

Figure 5.10: Comparison of C1s signal of SAM layers prepared in a nitrogen filled glovebox and under ambient conditions. When prepared under ambient conditions (right) the C1s peak structure shows a change between one hour and seven days immersion times, whereas the samples prepared under inert atmosphere does not.

5.3. Perfluorodecanethiol (PFDT) as Model System

using the extrapolation in the saturation region (ESR) method as described in [146, 148]. All above discussed effects like the influence of the molecular concentration in solution, the immersion time and the atmospheric condition on the charge injection functionality (corresponds directly to threshold voltage of OFETs) of the SAM are presented in Figure 5.11. Lines connecting the data points act as guide to the eyes. On the left side of this figure the threshold voltages of OFET devices with source and drain electrodes treated with PFDT SAM as a function of immersion time are plotted. For very low exposures the $|V_{th}|$ is higher than the $|V_{th}|$ of the untreated devices, which indicates a higher injection barrier. With increasing exposure the injection barrier decreases and the best device performance is reached for an exposure of $\sim 10\,\text{s} \cdot \text{mM}$. Ambient processed SAMs decrease in performance after longer immersion time (independent of concentration), whereas SAMs processed in nitrogen atmosphere do not show this effect. The initial change in performance is clearly connected to the exposure and surface coverage, while the final increase of $|V_{th}|$ and hence the performance decrease is only related to the immersion time and not to the concentration (see Figure 5.11 right). This confirms the suspicion that the decrease in SAM functionality is most probably related to the chemical reaction with the environment.

5.3.3. Discussion

In summary, several samples treated with different SAM concentrations were characterized with various analytic techniques as well as investigated within OFET devices. The experiments were always carried out in two different environment conditions: one sample preparation was continued in nitrogen atmosphere and the other one in ambient clean room conditions. In case of in detail studied PFDT SAM a WF shift of $\sim 0.5\,\text{eV}$ was reached, which is in very good agreement with values reported in the literature [153]. Moreover, such a significant shift was also reached on substrates dip coated in PFDT solution within only a few seconds. The same applies to the OFET devices which show increased performance due to the enhanced hole injection into a p-type organic semiconductor. The average angle with respect to the surface normal derived from IRRAS measurements for a complete PFDT monolayer was found to be $10° \pm 3°$. Furthermore, oxygen contamination leads to loss of

5. Self-Assembled Monolayers

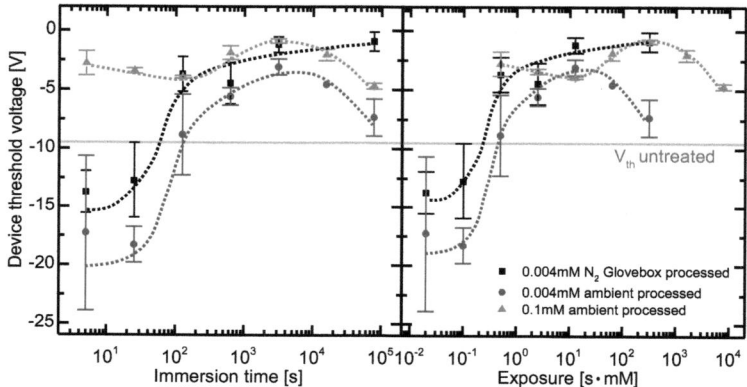

Figure 5.11: V_{th} of OFET devices with source and drain electrodes treated with PFDT SAM as a function of immersion time (left). V_{th} of OFET devices as a function of molecular exposure to the PFDT SAM (right). Each data point is an average value for at least 4 devices with the standard deviation shown as statistical error. Lines connecting the data points are guide to the eyes. Representative selection of chosen OFET characteristics before and after PFDT treatment is presented in Appendix A Figure A.4.

5.3. Perfluorodecanethiol (PFDT) as Model System

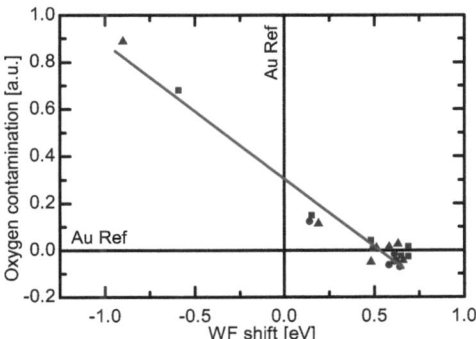

Figure 5.12: Oxygen contamination of ambient processed SAMs against WF shift.

PFDT functionality and in extreme cases results even in a WF shift in the opposite direction. In Figure 5.12 a compilation of the results achieved on ambient processed SAMs is presented. The WF shift is plotted against the oxygen contamination. The results show a clear trend that higher oxygen contamination leads to decrease in PFDT functionality and as shown in Section 5.2 the problem seems not to be only limited to this particular SAM molecule. However, it occurs only after several hours of immersion and only in ambient conditions. Therefore, if high enough concentration is chosen in order to complete the self-assembly process within short period of time this effect should not restrain of SAM injection layers.

In Figure 5.13 a compilation of all performed measurements on PFDT monolayer prepared in inert glovebox atmosphere is presented. Each data set exhibits very similar behaviors and indicates the end of monolayer formation at about $\sim 10\,\text{s} \cdot \text{mM}$ exposure. The F1s emission line intensity which acts as indicator for surface coverage is fitted to a Langmuir isotherm adsorption behavior (green line). Additionally, it can be used as guide to the eyes to follow the other data points.

5. Self-Assembled Monolayers

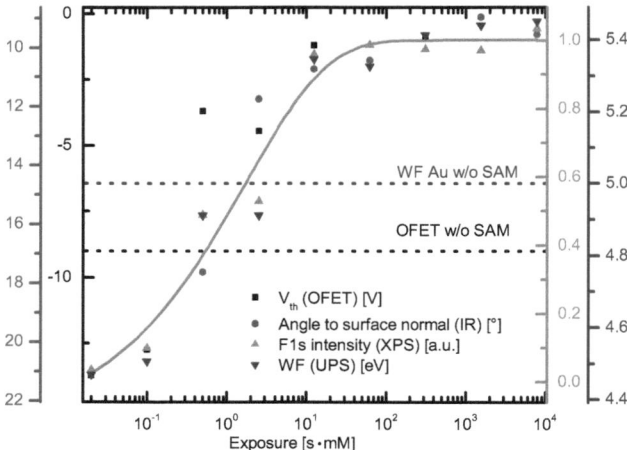

Figure 5.13: Compilation of measurements performed in this work in collaboration with Sabina Hillebrandt (IR) and Milan Alt (OFETs). The F1s emission, as indicator for surface coverage, is fitted to a Langmuir isotherm adsorption behavior (green line). All data shown in this figure was derived from SAMs prepared in an inert glovebox atmosphere.

5.4. Novel SAMs

One goal of the MORPHEUS project was to synthesize new SAM molecules which are not only limited to tune the electrodes WF, but also improve the wettability of the used substrate. Malte Jesper from OCI (University of Heidelberg) synthesized various SAM molecules. First, molecules with thiol anchor group were synthesized. Afterward, the same concept was used for SAMs with phosphonic acid as an anchor group. A full list of molecules synthesized by Malte Jesper is described in [163].

In this work three thiol and two phosphonate molecules were investigated which were the most stable and promising SAMs. The chemical structures of these SAMs are presented in Figure 5.14 and 5.15. All SAMs have electron donating substituents and thus also a negative dipole moment (along their axis in the direction of the anchor group). The quantification of the expected

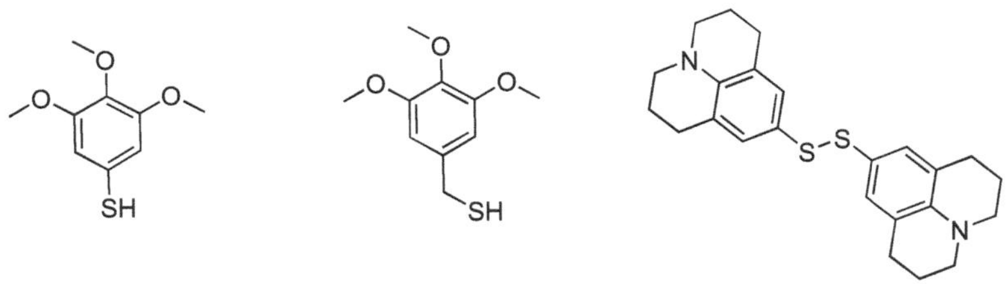

5.4. Novel SAMs

3,4,5-trimethoxyphenylthiol	3,4,5-trimethoxybenzenethiol	Bisjulolidyl disulfide
TMP-SH	TMB-SH	Juls

Figure 5.14: Chemical structures of new synthesized SAM molecules with thiol anchor groups characterized within this work.

Property	TMB-SH1	TMB-SH2	TMP-SH	Juls
Tilt angle [°]	18	73	28	8
Calc. WF shift [eV]	−0.31	−0.12	−0.43	−1.3
Dipole moment μ [D]	1.35	1.35	1.48	4.20
Dipole moment $\mu_\perp$ [D]	0.97	0.64	1.04	3.91
Preferred packing structure	$2\sqrt{3} \times 3$	3×3	$2\sqrt{3} \times 3$	2×2

Table 5.1: Calculated properties of single SAM molecule on Au(111) surface. Calculations performed by Iva Angelova from BASF SE.

dipole moment is presented in Table[1] 5.1. However, it must be noted that the calculated values should be regarded as guidelines rather than absolute values (only available for SAMs with thiol anchor group). The respective dipole moments were calculated for the free thiols. All quantum chemical calculations were performed by Iva Angelova (PhD student at BASF SE).

5.4.1. Aromatic Thiols on Gold

TMP-SH and TMB-SH are very similar, the only difference is the additional methylene linker for TMB-SH. This provides the molecule with a greater

---

[1] In case of TMB two calculations are made as the possibility exists that the molecule is "lying" on the surface.

5. Self-Assembled Monolayers

(3,4,5-trimethoxyphenyl)phosphonic acid
TMP-PA

(3,4,5-trimethoxybenzyl)phosphonic acid
TMB-PA

Figure 5.15: Chemical structures of new synthesized SAM molecules with phosphonic acid as an anchor group investigated within this work.

freedom in the orientation. According to the calculations the thiolate-gold bond of TMP-SH has an angle of about 28° to the vertical. Due to the additional methylene linker the TMB-SH is not limited to the standard bond but it can orient the functional group either perpendicular or parallel to the surface. Depending on the orientation of the benzene ring the direction of the dipole is also changed. The possible orientations of all used thiols are presented in Figure 5.16. This cartoon also shows the dipole in the direction of the surface, even though the oxygen atoms are highly electronegative. This is due to mesomeric effects. The free electron pairs of the oxygen electrons are pushed into the electron system of the benzene ring. The electron density increases and an activated benzene ring is obtained which is responsible for the occurring dipole orientation [164]. The contribution to the surface dipole and hence the change in the WF is greater if the deviation from the vertical alignment is smaller. The same applies to the Juls SAM.

All prepared monolayers were first measured with goniometer setup and KP in order to establish the optimal preparation process for the new molecules. The gold substrates were prepared and cleaned according to the methodology described in Section 3.2.1. One has to mention that these measurements were performed before the findings which have been obtained during the PFDT investigation. The gold substrates were immersed for 24 hours. The KP results are presented in Figure 5.17. TMP-SH and TMB-SH show as expected very similar WF shifts, which also corresponds well to the above presented

5.4. Novel SAMs

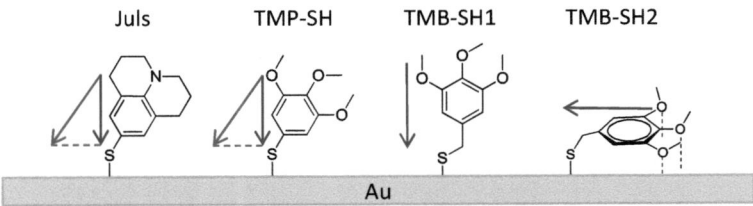

Figure 5.16: Possible orientations of the investigated SAM molecules and the corresponding effective dipole marked with red arrow.

calculations. The shift of TMB-SH is around ~ 250 meV and thus is between the calculated values for different orientations. The highest WF shift was achieved with Juls SAM and it was around ~ 1 eV. As the KP technique is a non destructive method, the same samples were than measured with the goniometer. The contact angle determinations were carried out as static contact angle on a Drop Shape Analyzer from Krüss (DSA100) with a drop size of 7.5 µL. At least two different solvents are needed in order to calculate the surface energy and to derive the wetting envelope, which describes the wetting properties of the investigated sample [165–168]. Four solvents were used to perform the contact angle measurements: water, diiodomethylene, glycerol and ethylene glycol. The wetting envelopes were calculated with "OWRK"[2] method [169–171]. The results are depicted in Figure 5.18. Compared to the bare gold surface, TMP-SH and TMB-SH only show little impact on the wetting envelope. Interestingly, TMB-SH shows a significantly larger impact on the dispersive factor than TMP-SH. This could be due to the increased degree of freedom imposed by adding the methylene linker. The methylene linker allows the molecule to stand perpendicular to the surface, but at the same time allows for flat lying molecules. The π-systems of those molecules would explain for the impact on the dispersive factor. The highest impact on the wetting envelope showed Juls almost doubling the covered area. The expanding of the wetting envelope by applying the new SAM molecules allows for a wider range of possible solvents that could be used to deposit further

---

[2] Owens, Wendt, Rabel and Kaelble (1969,70) - short "OWRK" allows, for at least two contact angle data, a graphical solution method.

5. Self-Assembled Monolayers

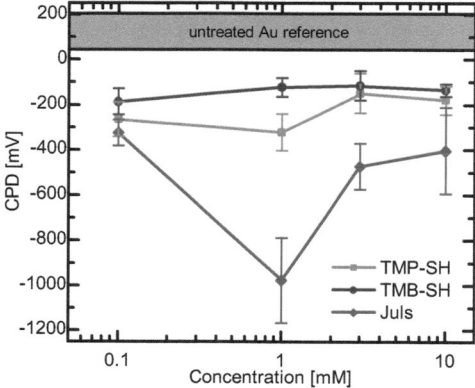

Figure 5.17: KP measurements of TMP-SH (green), TMB-SH (blue) and Juls (red) treated gold substrates. The gold substrates were immersed in various concentrations (ethanol was used as a solvent). Each data point is an average of at least 4 samples measured on two different point to ensure layer homogeneity.

layers on top.

5.4.1.1. Properties and Different Growth Mechanisms of the New Molecules - PES and IR

As a next step PES and IR measurements were performed. Several samples were prepared and different exposures (see Section 5.3) were used in order to monitor the growth mechanisms and to obtain the tilt angle θ of the completed monolayer. All PES measurements were performed by Valentina Rohnacher (TUD), former bachelor student. In Figure 5.19 the survey spectra of clean gold and gold treated with the three different SAMs are presented. All survey spectra are easily assigned to the investigated SAM molecules. In case of Juls treated substrate no nitrogen is visible at this moment. This is due to low cross section of this element. In Figure 5.20 fits of C1s core level spectra of TMP-SH (left) and TMB-SH (right) are presented. Black dots correspond to the measured spectra with removed background. The red line shows the resulting fit. In both graphs an error curve is plotted (the error curve is 5

5.4. Novel SAMs

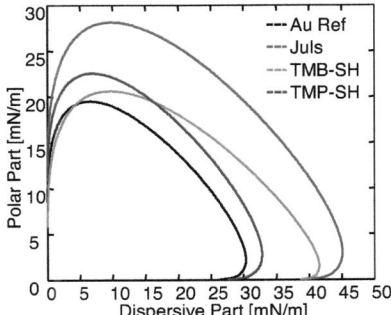

Figure 5.18: Wetting envelopes measured on gold and gold treated with a given SAM molecule. All contact angle measurements were performed with drop shape analyzer from Krüss (DSA 100). For more information about the measurement technique see [172].

times magnified). The C1s emission line shows two different peaks which can be assigned to the two carbon species within the TMB-SH and TMP-SH molecule. Due to the very high surface sensitivity of XPS measurements even the small difference between these two molecules is visible (additional methylene linker). In both cases the blue marked peak corresponds to the carbon atoms bound to oxygen. These atoms have the highest oxidation state within this molecule and as expected are chemically shifted to higher binding energies. A precise assignment to values from literature is not possible since the displacement is not only dependent on the C-O bond but is also influenced by the benzene ring. The same applies to the carbons marked with green color. In case of Juls SAM as expected no such behavior was observed due to its simple chemical structure (from the XPS view).

In the next step an exposure experiment (same approach as described in Section 5.3) was performed in order to investigate all three SAM molecules. All samples were prepared in the same manner as described in Section 3.2.1. The whole preparation process was carried out in inert glovebox atmosphere in order to avoid any oxygen and humidity related problems. Furthermore anhydrous ethanol was used as solvent. Very low concentration of 0.004 mM (just like in case of PFDT) of the given SAM was chosen for the used solution

5. Self-Assembled Monolayers

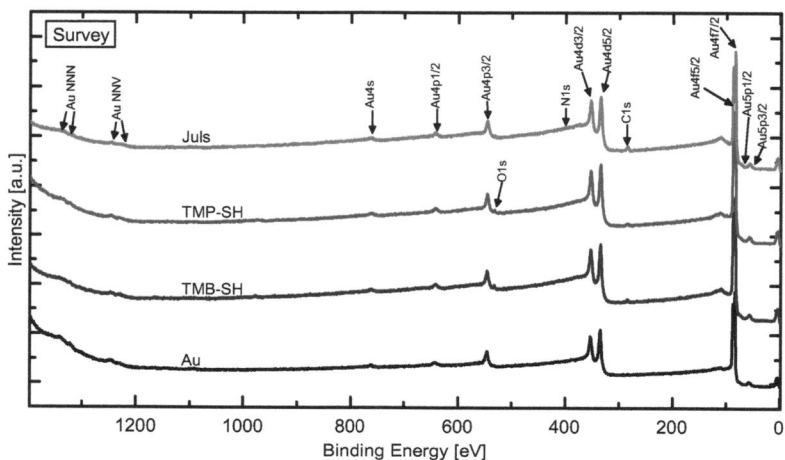

Figure 5.19: Survey spectra of gold and gold treated with TMB-SH, TMP-SH and Juls SAMs. All survey spectra are easily assigned to the investigated SAM molecules.

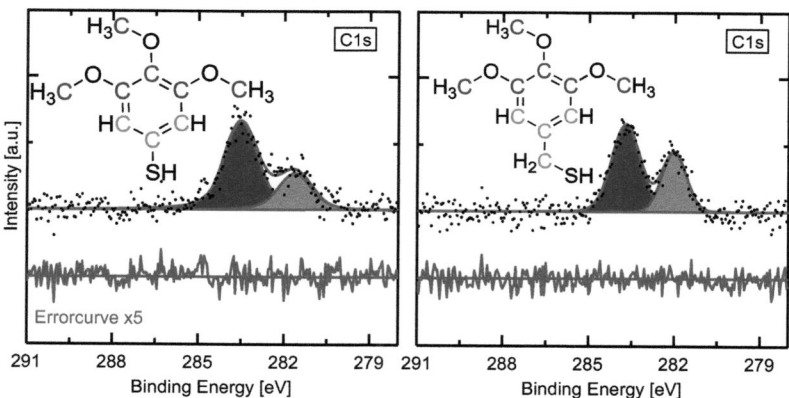

Figure 5.20: Assignment of C1s emission lines to the molecular structure of TMP-SH (left) and TMB-SH SAM (right).

5.4. Novel SAMs

with the hope that the growth process of the monolayer would be visible. It turned out that the used concentration was way too low. Even after 24 hours almost no SAM was detected on the investigated samples. Several attempts were taken in order to find out the right concentration or to adjust the exposure, unfortunately without the expected effect. It seems that for these SAMs the region between the no monolayer formation and fast formation is very narrow. Due to limited amount of available material no further attempts were taken in order to monitor the growth process. Nevertheless some important insights were gained. In Figures 5.21, 5.22 and 5.23 a representative selection of PES measurements are presented. In Figure 5.21 the C1s, O1s detail spectra as well as SE cutoff of substrates treated with TMP-SH SAM are shown. Only two immersion times are shown, the shortest and the longest one as no noteworthy changes were obtained in between. In this case a concentration of 1 mM was used. Already after a five seconds immersion time the same amount of molecules on the surface as well as a very similar WF shift is measured as for a 24 hours immersion time. A WF shift of around $-0.6\,\text{eV}$ was measured for the very short immersion time. This result is higher than the values previously measured with KP. However, this relative small difference can be explained by the optimized SAM process as well as the fact that using UPS the smallest WF within the measuring spot is always measured whereas in case of KP measurements it is always an average value. For the same set of samples IRRAS measurements were performed in order to determine the final tilt angle θ of the completed monolayer. The measurements were performed and evaluated by Sabina Hillerbrandt and Joshua Kreß (University of Heidelberg). The resulting tilt angle is around $\sim 18° \pm 2°$. This upright angle could also partly explain the difference between the calculated WF shift $\Delta\phi = -0.43\,\text{eV}$ and the measured value $\Delta\phi = -0.6\,\text{eV}$. However, this determined angle should not be over interpreted as for this sample set some problems with intensities occurred. For more details and measured IRRAS spectra see [173].

The TMB-SH treated substrates show a very similar behavior (see Figure 5.22). Again only shortest and longest immersion times are presented as no noteworthy changes in between were observable. A concentration of 0.1 mM was used in this case. No differences in the amount of molecules between these two exposures is detectable. Only a minor change in the WF is obtained. The

5. Self-Assembled Monolayers

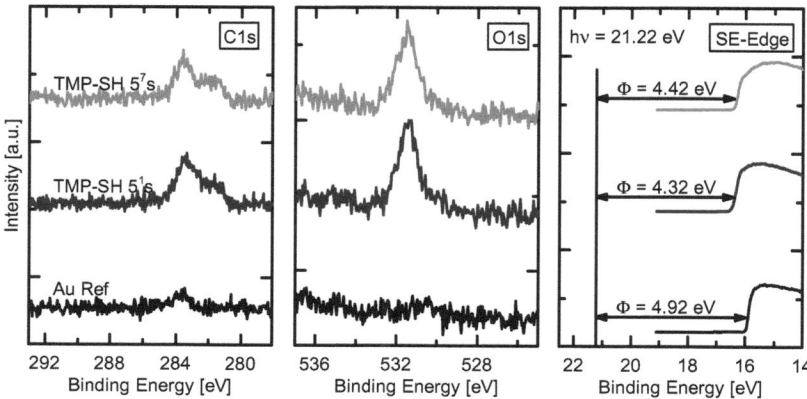

Figure 5.21: Representative selection of PES measurements on pure gold and gold treated with TMP-SH SAM. From left to right detail spectra of C1s, O1s core level and SE cuttoff with the absolute WF values.

maximum WF shift measured using the TMB-SH SAM was $\Delta\phi = -0.43\,\text{eV}$. The UPS measured shift is again higher than the one previously determined with KP. The explanation is however similar to the above. One has to mention that the reached shift was maybe reduced by the fact that the used gold reference was slightly contaminated with some adsorbates, although the substrate preparation was carried out in the same manner. This contamination however is not hindering the TMB-SH molecule to self assembly on the gold surface. Already after five seconds immersion time the initial contamination is replaced by the used SAM. However, an overall influence of the contamination on the SAM monolayer cannot be excluded. In collaboration with Sabina Hillerbrandt and Joshua Kreß IRRAS measurements were performed. The spectra are not shown here but can be found in [173]. In this case, no problems during the measurements occurred and the two samples sets show very similar results. The measurements show that initially the tilt angle of the TMB-SH after five seconds immersion time is very high ($\theta = 56° \pm 6°$) which means that the molecule is more "lying" on the surface than the TMP-SH. As the immersion time increases the angle is getting even higher. After 48 hours immersion time the tilt angle reaches $\theta = 62° \pm 6°$, which means that the

5.4. Novel SAMs

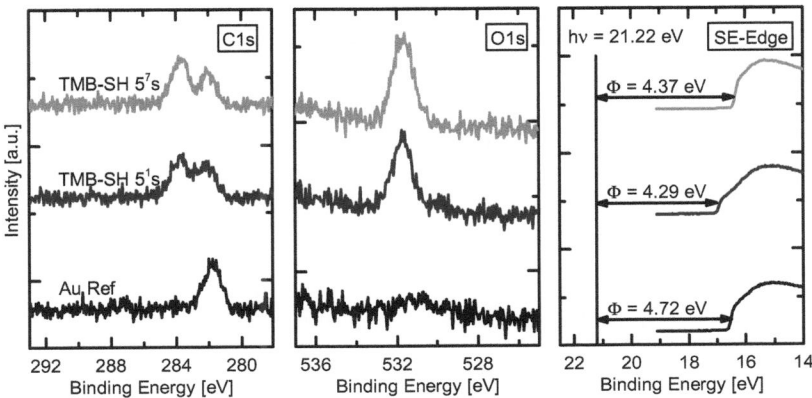

Figure 5.22: Representative selection of PES measurements on pure gold and gold treated with TMB-SH SAM. From left to right detail spectra of C1s, O1s core level and SE cuttoff with the absolute WF values.

molecule lies down even more on the surface than before. It seems like the additional methylene linker leads to the change in the tilt angle, however not in the desired direction. Taking this into account the achieved WF shift is remarkably high (see Table 5.1).

Finally the Juls SAM was investigated which trough the KP and goniometer preliminary investigation showed superior functionality. Representative selection of PES measurements is presented in Figure 5.23. A concentration of 0.02 mM and 1 mM was used. In this case data for three different immersion times is shown. After five seconds immersion time an intensity increase in the C1s detail spectra is clearly visible. For this low exposure a WF shift of $\Delta\phi = -0.68$ eV is achieved and is already higher than in case of TMP-SH and TMB-SH SAM. After 5^4 s immersion time small change in C1s detail spectra is obtained. Most probably at this point the surface is almost completely covered with Juls molecule and the self assembly process is completed. The measured WF of $\phi = 3.87$ eV confirms this impression. After almost 24 hours of immersion the measured spectra show almost no changes. Only the WF is slightly higher. The measured WF shift is remarkable and an influence of solvent or other adsorbates can be excluded, however still no nitrogen

5. Self-Assembled Monolayers

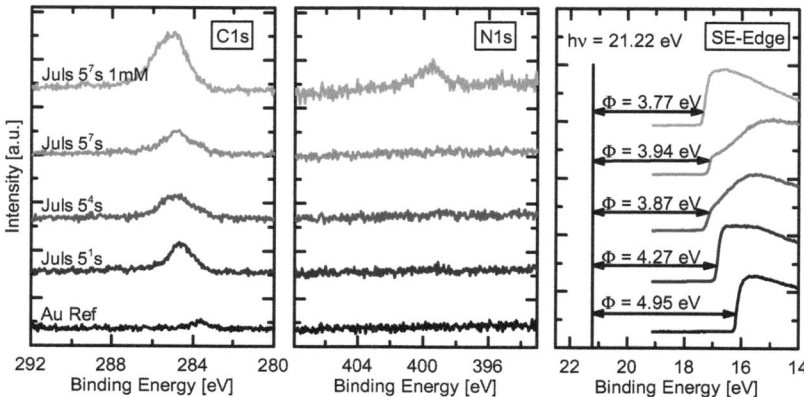

Figure 5.23: Representative selection of PES measurements on pure gold and gold treated with Juls SAM. From left to right detail spectra of C1s, N1s core level and SE cuttoff with the absolute WF values.

was detected. The last measurement (turquoise) was performed on a sample immersed in solution with much higher Juls concentration (1 mM). Both detail spectra show an increase in signal intensity, however the intensity of the carbon peak is very high. Various reasons can explain such a behavior, either the SAM molecule is densely packed or a multilayer is formed. Albeit the first one seems to be the case as the SE cutoff shows even higher WF shift. The achieved WF shift is higher than previously measured with KP which again point out the differences in the measurements methods. Also in this case IRRAS measurements were performed. However, various technical problems and contamination of the whole sample set prevented obtaining reliable data. The WF shift calculated by Iva Angelova for this molecule was $\Delta\phi = -1.3\,\text{eV}$ and is marginally higher than the measured one ($\Delta\phi = -1.18\,\text{eV}$). One possible reason could be that the tilt angle is responsible for this difference.

5.4.1.2. OFET Devices

Due to the satisfactory results mentioned above OFET devices were prepared featuring those new synthesized SAM molecules. All OFET devices presented in this study were prepared and evaluated by Milan Alt (KIT). The device

5.4. Novel SAMs

layout used for this purpose is described in Section 3.3.2.2. One has to mention that in this case the investigated SAMs act as interfacial layer in order to improve the electron injection in the N2200 [174] (for chemical structure see Appendix A, Figure A.3) semiconductor layer. Transfer characteristics were measured in the linear regime. The results are summarized in Figure 5.24. A general improvement can be observed when introducing the SAM as additional functional layer. The TMP-SH and TMB-SH show similar impact on the transistor characteristics, increasing saturation mobility and on/off ratio. This is surprising as major differences between those two substances in all other performed measurements were observed. Assuming the partially formed SAM with flat laying moieties as the surface in the TMB-SH case, it is possible that the deposition of another layer on top interacts with the flat lying molecules causing them to stand up. The so formed layer would be a little less dense than the one formed from the TMP-SH, but at the same time having larger effective dipole moment. This would explain the differences in former measurements as well as the similarity in device characteristics. The strongest impact on OFET performance has Juls, as is expected by its strong intramolecular dipole and therefore strong WF shift. The same figure shows the derived contact and channel resistance via transfer line method (TLM) [175]. The contact resistance was reduced nearly by two orders of magnitude from $8 \cdot 10^{-6}$ to $1.4 \cdot 10^{-5}$ Ωcm in case of OFETs treated with Juls. TMB-SH and TMP-SH SAMs reduce the contact resistance by a factor of two to $\sim 4 \cdot 10^{-6}$ Ωcm with respect to the untreated Au contacts. The derived channel resistance (Figure 5.24, graph in the middle) shows almost no changes between the untreated and treated samples. Only the TMP-SH treated devices show a small decrease in the channel resistance which is most probably related to some problems with the model used for the calculations.

To compare the functionality of Juls OFET devices with PEIE (which acts as a benchmark system for lowering the WF, see Section 4) treated electrodes were prepared by Milan Alt according to the process described above. The results are presented in Figure 5.25. One has to mention that in this case silver was used as electrode material which is a common used drain/source electrode. Again, the OFET characteristics show an improvement of threshold voltage if an interfacial layer was used. Channel and contact resistance derived via TLM

5. Self-Assembled Monolayers

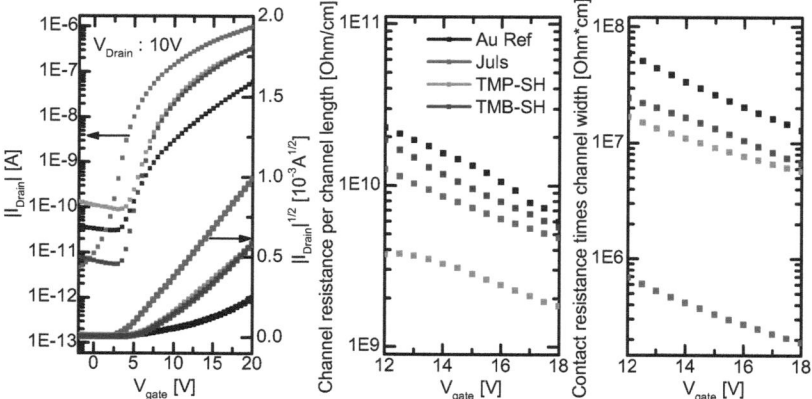

Figure 5.24: Left: OFET transfer characteristics without SAM treatment (black), with TMP-SH treatment (green), TMB-SH treatment (blue) and after Juls treatment (red). The applied V_{SD} is 10 V. Middle and right: channel resistance per channel length and contact resistance times channel width, respectively, both derived from TLM. All data shown in this study are average values from at least 5 devices.

method confirm this impression. While the channel resistance remains stable for untreated and treated devices, the contact resistance reduces by almost one order of magnitude for PEIE treated OFETs with respect to the untreated electrodes. Juls SAM treated OFETs show only half as large improvement. No stack optimization were taken in order to maximize the achieved results, which should be done in near future as the Juls treatment process was optimized for gold electrodes. Taking this into account an improvement of Juls functionality as an interfacial layer in OFET devices with silver electrodes is in reach.

5.4.2. Aromatic Phosphonic Acids on ITO

A transparent electrode, in particular ITO, is crucial for organic electronic devices like OLEDs or OSCs, thus controlling the surface properties of ITO is of great importance. In order to transfer the same SAM concept presented above to metal oxides the anchor group of these molecules needs to be changed as S-H group is not favored for binding to ITO surface [176–178]. To bind with

5.4. Novel SAMs

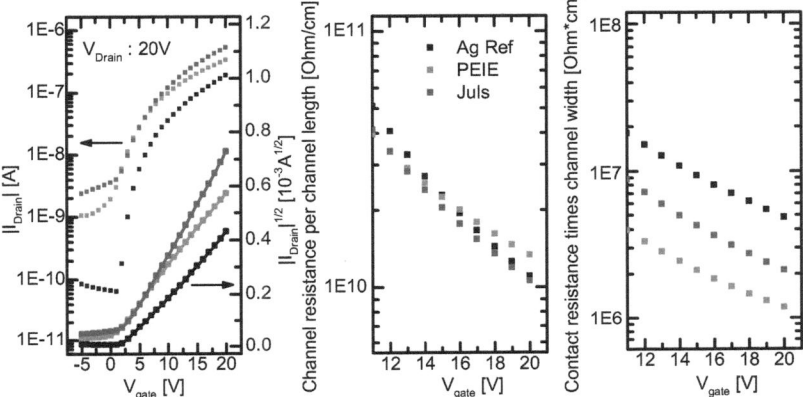

Figure 5.25: Left: OFET transfer characteristics without SAM treatment (black), with Juls treatment (red) and with PEIE treatment (green). The applied V_{SD} is 20 V. Middle and right: channel resistance per channel length and contact resistance times channel width, respectively, both derived from TLM. All data shown in this study are average values from at least 4 devices.

ITO carboxylic and phosphonic acids are preferred. However, phosphates and phosphonates show well ordered SAMs on ITO and moreover it is reported that their bond to metal oxides is stronger than in case of carboxylic acids [179–182]. Malte Jesper (OCI, University of Heidelberg) synthesized the counterpart to previously investigated SAMs with phosphonic acids as anchor group. Unfortunately the Juls-PA proved to be unstable. The chemical structures of TMP-PA and TMB-PA are shown in Figure 5.15. Both SAM molecules were measured with KP in order to establish the optimal preparation process. The ITO substrates were prepared and cleaned according to the recipe described in Section 3.2.1. As the binding mechanisms of phosphonic acids differ from those of thiol group and metal, different preparation processes were used. For more detail about the additional preparation steps see [183]. Chokalingam et al. [96] reports about strong influence on the quality of SAM formed from phosphonic acid depending on characteristics of used ITO substrate. Smooth amorphous ITO leads to monolayers with significantly lower amount of defects within the monolayer than in case of polycrystalline ITO. Both types of ITO

5. Self-Assembled Monolayers

were investigated. The KP results are presented in Figure 5.26. As expected the TMP-PA and TMB-PA show similar results. The shift of TMP-PA is around $\sim 1\,\text{eV}$ with respect to oxygen plasma treated ITO substrate and thus it is around $\sim 100\,\text{meV}$ higher than in case of TMB-PA. The very high WF shift should not be over interpreted as the ITO reference has very high WF due to the plasma treatment (around $5.5\,\text{eV}$ in comparison to around $4.6\,\text{eV}$ before plasma treatment), however it was the only way to have a relative stable ITO value for a comparison. For this purpose HOPG was also measured in order to obtain the absolute WF of the treated substrates (right graph, an assumption was taken that HOPG has a WF of $\phi \approx 4.47\,\text{eV}$). The absolute WF of the substrate treated with TMP-PA is around $\phi \approx 4.4\,\text{eV}$ and accordingly $\phi \approx 4.5\,\text{eV}$ for TMB-PA treated one. These values were achieved on substrates with polycrystalline ITO (TMP_p and TMB_p). On amorphous ITO $\sim 200\,\text{meV}$ higher WF shift could be achieved (TMP_a and TMB_a). TMP-PA and TMB-PA provided a WF of $\phi \approx 4.18\,\text{eV}$ and $\phi \approx 4.19\,\text{eV}$, respectively. The very similar values provided by those SAMs on amorphous ITO are most probably a hint on more dense monolayer with less defects which confirms the idea reported in [96]. In this work no further measurements for these molecules were performed. For further investigations and device characteristics with TMP-PA and TMB-PA as interfacial layer see bachelor work by Ramos-David Bwalya (University of Heidelberg) [183].

5.5. Conclusion

In this work various SAM molecules on different substrates were investigated. At the beginning some well known and commercially available SAM molecules were characterized in order to establish the process at InnovationLab. During this analysis new insights were found. It was proven that a deposition of SAM from solution in line with a high throughput printing process is in principle possible. However for successful application an optimization of crucial process parameters is of great importance. One needs to ensure an exposure (concentration × immersion time) above the critical value concerning full coverage and if processed under ambient atmosphere conditions, immersion times short enough to avoid degradation. Furthermore each contamination of

5.5. Conclusion

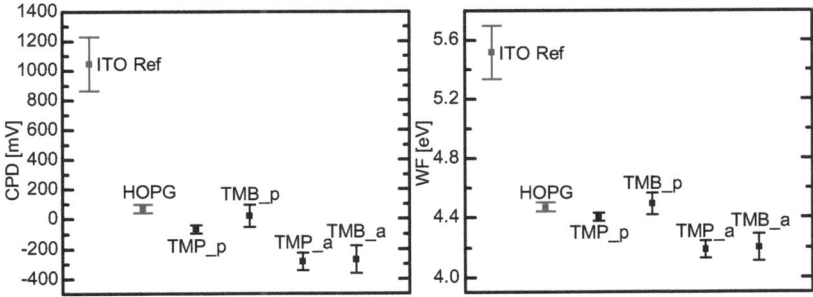

Figure 5.26: KP measurements of ITO reference directly after oxygen plasma treatment and ITO treated with TMP-PA and TMB-PA SAM molecules. The change in the WF is relative to an untreated reference sample. Each point represents the average value obtained from at least 4 samples. Polycrystalline (_p) and amorphous (_a) ITO was used as a substrate. Additionally HOPG measurements were performed.

the substrate is not necessarily hindering the functionality of the monolayer but it has a large influence on how fast the self assembly process takes place.

In the next step new molecules, synthesized by Malte Jesper (OCI, University of Heidelberg), were investigated in detail. KP and goniometer measurements were first performed in order to find the optimal process parameters. Already this preliminary investigation provides promising results, with WF shift of around $\sim 1\,\text{eV}$ and almost doubling the area of wetting envelope was achieved on substrates treated with Juls SAM. Using the new insights from PFDT investigation, similar characterizations were performed on the new molecules using PES and IRRAS techniques, which confirmed and showed even higher WF shifts for each molecule. Additionally in collaboration with Milan Alt (KIT) several batches of OFET devices were prepared and characterized. The contact resistance derived by TLM method shows nearly two orders of magnitude lower values (from $8 \cdot 10^{-6}$ to $1.4 \cdot 10^{-5}\,\Omega\text{cm}$) in case of OFETs treated with Juls. Moreover a very first comparison between PEIE benchmark system against Juls provides promising results. First preliminary results for the same SAM concept but adapted for metal oxides were obtained and work

5. Self-Assembled Monolayers

function shifts by more than 1 eV were achieved.

Putting all these results together a new class of SAMs was derived to be used in organic electronics. It consists of small molecules and therefore allows for unhindered charge transport across the interface and is able to significantly change the work function of noble metals. It greatly improves the wetting behavior of metal surfaces and most probably of metal oxides allowing for a wider range of solvents being used in follow-up steps. It could be shown that these SAMs are capable of significantly improving device characteristics of an OFET device. Emerging from the success of these rather simple molecules, one can imagine a whole lot more future SAM molecules. Molecules with even higher dipole moments can be synthesized, having even a higher impact on the work function. Alternative synthetic approaches would allow for further functionalization of the molecules. This would give access to impart crystallization dominating moieties while at the same time controlling the electric (work function) characteristics.

6. Organic Semiconductors with thermally activated solubility reduction

Torben Adermann from OCI (University of Heidelberg) synthesized an organic semiconductor which can be processed from solution, and through a pyrolysis process the solubilizing groups are removed and the semiconductor becomes insoluble again. Moreover, after the whole process the semiconductor should still have an n-type behavior. If possible it should work in the same manner for polymers as well as for small molecules and the pyrolysis temperature should not exceed 200 °C, in order to be able to use also flexible substrates. To achieve this goal the 1,4,5,8-naphthalenetetracarboxdiimide (NDI) was selected as origin material. Two approaches were used within this work, the first one with NDI-carbamates and the second one with NDI-linker-carbonates (see Figure 6.1). The influence of these methods on thin layers processed from solution with small molecules as well as with polymers are discussed in detail within this work. All synthesis steps and a more comprehensive investigation in a chemical manner can be found in the PhD thesis of Torben Adermann [184]. In the first part of this chapter a short overview about recent work and results of other groups working in the field of organic semiconductors and solution processing will be given. The next part summarizes the results achieved using small molecules and polymers. FTIR, PES, AFM measurements as well as devices are introduced. Finally, a comprehensive comparison and discussion considering all experimental findings is given.

6. Organic Semiconductors with thermally activated solubility reduction

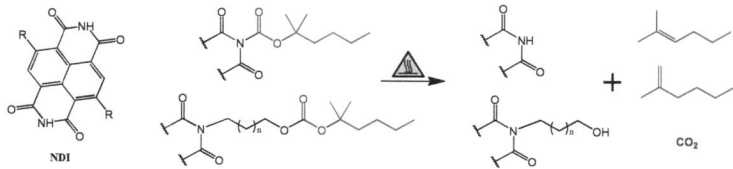

Figure 6.1: Chemical structure of the origin material (left) and the two different approaches which were used in this work. Thermal cleavage of NDI-carbamates (top) and NDI-linker-carbonates (bottom). Taken from [184].

6.1. State of the art

There are various possibilities to manufacture insoluble layers of organic semiconductors. One of them is by evaporation of small molecules in vacuum conditions, which is so far the most common method used in the industry. This method is widely used to build OLED displays for small electronic devices like cell phones or tablets. However, this prevents the possibility to tap the full potential out of organic electronics, which is large area production, because vapor deposition is very limited and expensive. To achieve large area production materials are needed which can be litigated from the liquid phase. However, like already mentioned in Section 1.1, this can lead to intermixture of materials and the number of available orthogonal solvents to work around the problem is limited. For this reason materials are needed which not only can be processed from solution but also with a specific material transformation are able to loose their solubility. In the literature, two methods are proposed to achieve this result: the covalent cross-linking and cleavage of solubilizing groups.

Using the covalent cross-linking method it is possible to produce multilayer OLEDs with the help of various cross-linked and with oxetane functionalized semiconductors [185–187]. Because of two major disadvantages of this approach, it is not going to be a topic in this work. Indeed, in this process, even though the external thermal stimulus is used, polar reagents for the cross-linking reaction are also needed, which act as a potential hazard for the electron transport for n-type semiconductors (which are the topic within the

6.1. State of the art

scope of this work). Moreover, a covalent cross-linking leads to irreversible fixed morphology and cannot be optimized with further thermal treatments. This work addresses the cleavage of solubilizing side chains where through the cooperation with the synthesis group form OCI at University of Heidelberg, more precisely Torben Adermann, many new compounds could be tested as a thin film. A detailed description about the current development in thermal cleavage of solubilizing side chains in organic semiconductors and their application can be found in the review article by Krebs et al. [8].

6.1.1. Small Molecules

Back in 1999 Herwig and Müllen [188] presented a concept for how to thermally decompose a precursor molecule into insoluble pentacene and achieve a mobility of $\mu = 0.1$ cm²/Vs within a p-type transistor device. The next approach on how to remove the solubilizing groups was introduced by Ciba Specialty Chemicals corporation, where pigments where transferred into a soluble precursor by attaching carbamate[1] structures [189]. Afterwards, by thermal decomposition of the carbamates (from 145 °C) the insoluble material is obtained. The first successful attempts to transfer the thermally labile masking of aromatic amides to the materials for organic electronics were reported by Ma et al. in 2011 [190]. During this experiment a fully solution processed $Quinacridon : PC_{60}BM$ organic solar cell was created, which achieved an efficiency of 0.61% after the thermal treatment. Next attempt in 2013 by Glowacki et al. [191] delivered again a working solar cell. However, this time a small molecule with n-type behavior was used. This was the first report of a functional solution processed n-type semiconductor material, which was produced by thermal cleavage of a carbamate. The first OFET device processed with thermal labile masking of carbamates on DPP:Bitiophen oligomers was reported in 2012 by Yamashita et al. [192]. It is about ambipolar OFET devices prepared via solution shearing method, which after the pyrolysis show sufficient mobilities. So far no reports can be found on pure, air stable n-type semiconductor materials with such thermally cleavable chains. Within this work various NDI derivatives synthesized by Torben Adermann with low molecular weight are characterized.

---

[1] An organic compound derived from carbamic acid.

6. Organic Semiconductors with thermally activated solubility reduction

Further information about small molecules with thermal solubilizing groups and an insight from the chemical point of view can be found in [184].

6.1.2. Polymers

The first attempts on polymers with thermal cleavable side groups go back to 1980, where polyacetylene derivatives were investigated [193, 194]. In this concept, using the so called Retro-Diels-Alder reaction, the precursor polymer becomes polyacetylene due to thermal cleavage of volatile aromatics. Further experiments using this concept were also presented in [195]. The disadvantage of this method is obvious: the conjugated backbone of the polymer is not formed until the spin-off reaction is successfully done. If it is not the case then the conjugated backbone of the polymer is interrupted, which implies strong disadvantages for the electrical properties of the material. A possible solution to this problem can be to first synthesize the conjugated polymer, then add the solubilizing side chains and finally thermally cleave them after the desired layer is processed. Such an approach was presented by Holdcroft et al. in 2005 [196]. Another example with conjugated polymers and thermally removable side chains was introduced in [197]. This work describes polythiophene functionalized with solubilizing side chains of carboxylic acid esters. With a temperature of 200 °C a volatile non-polar alkene is cleaved. In order to remove the residual carboxylic acids, a temperature above 300 °C is needed. These polymers were successfully used by the Krebs group within organic solar cells, however not until the carboxylic acids were removed [198–201]. Pyrolysis of carbamates is also possible for polymer systems and was shown in [202]. Due to the pyrolysis process, a side effect occurs and leads to inversion of dominant polarity in ambipolar polydiketopyrrolopyrrole systems.

6.2. NDI Derivatives - Investigation of Synthesized Small Molecules

Various NDI derivatives were synthesized by Torben Adermann and all of them were investigated on their solubility and processability as a thin film. The most representative and investigated materials as well as their chemical

6.2. NDI Derivatives - Investigation of Synthesized Small Molecules

structures are presented in Figure 6.2. All investigated structures and the details why which side group was chosen from the chemical point of view can be found in [184]. One has to mention that for each tested molecule, only a very limited amount of material was at disposal which strongly limited the ability to perform various experiments with many iterations. In Table 6.1 the theoretical values of weight loss during the pyrolysis process and the experimental values measured through thermogravimetric analysis[2] (TGA) on powder can be seen. In the second row "Pyrolysis Temp." the TGA parameters can be found. All TGA data shown in this work were measured by Torben Adermann (OCI, University of Heidelberg). The "Onset" is referred to temperature where the first visible signal change can be manually determined with the assumption that the thermally labile groups start to decompose. The expression "5%" is used in the literature and describes the point where the organic material looses 5% of its own weight. In this case a controlled cleavage is investigated, thus the parameter is only of limited importance, however always to be found by TGA measurements. The "Middle" describes the temperature where 50% of observed weight loss has been achieved and is calculated by the readout software. It is an important parameter, however the evaporation rate of the cleaved products can vary and thereby also the middle point temperature. With this knowledge, the first assumptions for the pyrolysis of the thin layers can be made, namely when the pyrolysis process should start and how fast the decomposition products evaporate from the layer.

6.2.1. Dewetting and Pyrolysis

An attempt was taken to process all synthesized small molecules from solution and then to perform the pyrolysis treatment. Unfortunately, not all of them were soluble enough to achieve a homogeneous layer (see Figure 6.3a) or like in other cases various dewetting issues occur during the process (Figure 6.3d). Prior to solution processing, all substrates were pretreated as

---

[2] TGA is a method of thermal analysis, where the changes in properties of investigated material are measured as a function of increasing temperature or as a function of time. In this work TGA was used in order to monitor the weight loss of investigated material using a heat ramp of 10 K per minute.

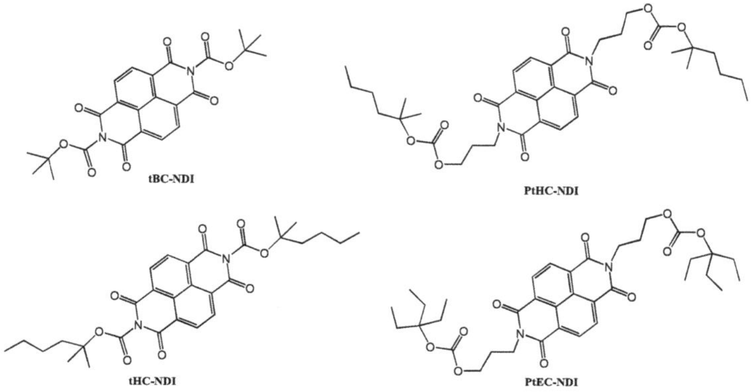

6. Organic Semiconductors with thermally activated solubility reduction

Figure 6.2: Chemical structures of chosen small molecule NDI derivatives.

Material/Properties		tBC-NDI	tHC-NDI	PtHC-NDI	PtEC-NDI
Weight loss	Calc	42.9%	51.6%	42.3%	42.7%
	Exp	42.0%	50.9%	41.7%	41.8%
Pyrolysis Temp.	Onset	165 °C	140 °C	170 °C	140 °C
	5%	198 °C	176 °C	206 °C	184 °C
	Middle	203 °C	185 °C	220 °C	198 °C

Table 6.1: The table shows the theoretical values of weight loss during the pyrolysis process and the experimental values measured through thermogravimetric analysis (TGA) on powder. Moreover, the list of TGA parameters can be also found.

6.2. NDI Derivatives - Investigation of Synthesized Small Molecules

described in Section 3.2.2.1. Additionally, for materials with dewetting problems further substrate treatments were performed like octadecyltrichlorosilane (OTS) treatment [203–205], hexamethyldisilazane (HMDS) treatment [206], various plasma treatments, ozone oven treatment as well as various series of substrate solvent treatment which should enhance the wetting properties [95]. Unfortunately, in most cases none of these attempts were very successful. The tBC-NDI for example delivered a very homogenous film on different substrates, however after the pyrolysis attempt was started only very inhomogeneous layers could be achieved (see Figure 6.3b, c, e). The processing of tHC-NDI and PtEC-NDI leads to very similar results. Due to the fact that none of these small molecules lead to successful results, they are not going to be discussed in detail within this work.

However, one has to mention that the PtHC-NDI showed a very interesting and maybe optimizable behavior. The first part of PtHC-NDI was dissolved in chlorobenzene (20 mg per ml) and spin coated on the substrate. Pyrolysis was performed in the previously described heating stage with the three temperature scenarios. All of them lead to the same result that can be seen in Figure 6.4. During the pyrolysis attempt, the 80 nm thick film is already starting to melt down at a temperature around 90 °C, and probably due to the surface energy is also dewetting. Various substrate pretreatments were investigated in order to overcome this dewetting issue. None of the standard pretreatments mentioned above provide an improvement, however it has been found that a substrate coated with PEI polymer could prevent the thin film from dewetting. Figure 6.5 shows the result of PEI pretreatment. The film spin coated on silicon substrate coated with PEI is able to survive the pyrolysis process. In the middle picture of Figure 6.5 a color change can be seen from orange to blue and at 207 °C a crystallization of the film can be observed. The dimension of the crystals is up to 100 µm. The thickness of the layer decreases almost to 30 nm (hard to measure because of the very rough surface). Further microscopy measurements were performed in order to determine if the resulting layer is homogenous enough for a transistor device. After some small optimization steps a satisfactory layer could be obtained (see Figure 6.6).

In collaboration with BASF SE temperature dependent X-ray diffraction (XRD) on the thin films was performed. Two samples were investigated, one

6. *Organic Semiconductors with thermally activated solubility reduction*

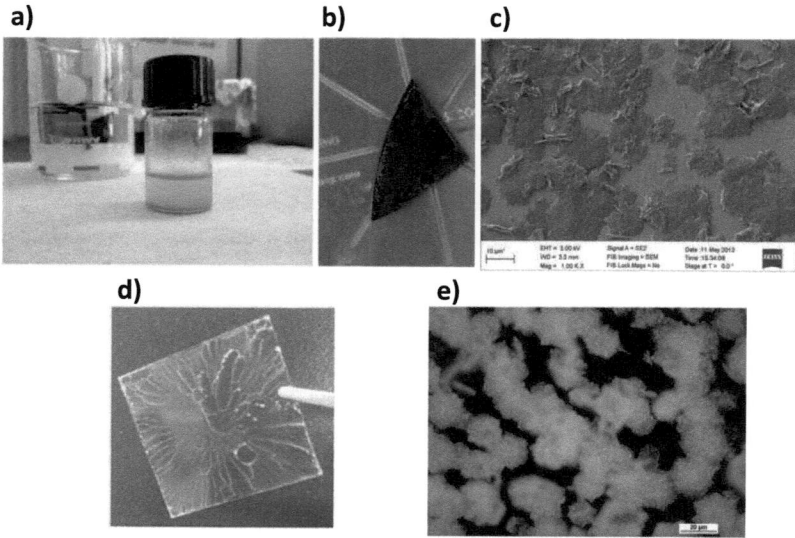

Figure 6.3: Representative selection of problems, which appeared by processing the synthesized small molecules and unfortunately could not be overcome within this work: a) the solubility issue, b) picture of not homogeneous layer after pyrolysis spin coated on a Si substrate, c) SEM picture of pyrolyzed layer - clearly recognizable islands of material on Si substrate, d) dewetting issues, e) no closed layer after the pyrolysis.

after the pyrolysis and an untreated one. The method confirms the melting point around 100 °C and the crystallization around 210 °C. Moreover, it shows significant differences between the crystal structure of already pyrolyzed sample and the sample held at 210 °C, which confirms the suspicion from [184], that during the cool down a phase transition occurs. Even after very intensive trials and many iterations, no functioning OFET devices with PtHC-NDI could be fabricated during this work, most probably due to the high roughness of the semiconductor or contact problems either with electrodes or with dielectric.

6.2. NDI Derivatives - Investigation of Synthesized Small Molecules

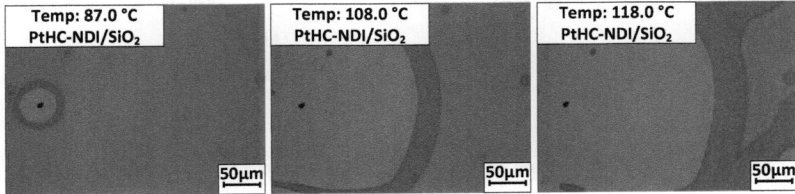

Figure 6.4: Screenshots taken from a video recorded with Olympus microscope with a heating stage. From left to right: pyrolysis attempt of PtHC-NDI on clean silicon substrate. Clearly a melting process can be seen which leads to dewetting of the previously spin coated material.

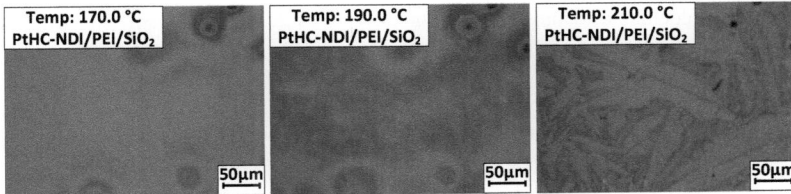

Figure 6.5: Screenshots taken from a video recorded with an Olympus microscope with a heating stage. From left to right: pyrolysis process of PtHC-NDI on a silicon substrate coated with PEI polymer. First, significant change in the color is visible and afterwards at around 207 °C the whole thin film crystallized.

6.2.2. Summary

Although, the studies with the new synthesized small molecules were not very successful yet, there are some further improvements or ideas which can be taken into account. It would be for example conceivable to structure the substrate pretreatment which might lead to the crystallization along the treated surface. Moreover, one could also think about localized heating, which specifies the direction of the crystallization. Very first attempts were carried out within this work but the amount of the available material was not sufficient enough for a comprehensive study. Even though the phase characteristics are very complex and the processing of NDI small molecules derivatives did not lead to the desired results, it still remains a promising method to achieve

6. Organic Semiconductors with thermally activated solubility reduction

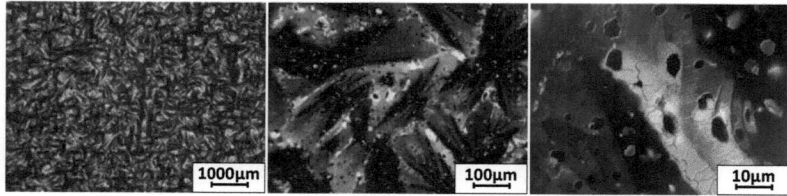

Figure 6.6: Polarized light microscopy images of the thin film of PtHC-NDI after pyrolysis process. The magnification is increased from left to right. Many defects and holes can be seen especially in the picture on the right side.

the goal of this project. Further details about the remaining investigated small molecules and more comprehensive considerations about the chemical processes which occur during the pyrolysis can be found in [184].

6.3. NDI Derivatives - Investigation of Synthesized Polymers

In this section new synthesized polymers are investigated. The methods known from previous sections are transferred to NDI:bithiophene copolymers as illustrated in Figure 6.7. The amount of solubilizing side chains added to the conjugated polymer backbone should be reduced throughout the pyrolysis process until the insolubility is achieved without strong change in the homogeneity of the thin layer. In this work, three different polymers were investigated. Their chemical structure will be always shown in the corresponding section. In Table 6.2 the important TGA parameters can be found. The first successfully synthesized polymer which met the requirements was P(HtHC-NDI-4HT2). The chemical structure of the soluble polymer can be seen on the left side of Figure 6.8. On the right side of the same figure a chemical structure after pyrolysis process is presented. In this work, two batches of this polymer, each with a different molecular weight, were available which led to different results. For this reason these two batches are presented in two different sections.

6.3. NDI Derivatives - Investigation of Synthesized Polymers

Material/ Properties		P(HtHC-NDI -4HT2)	P(tHC-NDI -4HT2)	P(HtODC-NDI -T2)
Weight loss	Calc	26.3%	32.3%	48.5%
	Exp	24.3%	28.9%	42.0%
Pyrolysis Temp.	Onset	157 °C	97 °C	173 °C
	5%	193 °C	132 °C	203 °C
	Middle	200 °C	146 °C	225 − 300 °C

Table 6.2: Theoretical values of weight loss during the pyrolysis process and experimental values measured through TGA on powder. The TGA parameters are also shown in the table.

6.3.1. P(HtHC-NDI-4HT2) → P(NDI-C6OH-4HT2) Molecular Weight of 130 kDa

The first tests with TGA and gel permeation chromatography[3] (GPC) performed by Torben Adermann (see [184]) promised a good solubility in chlorobenzene, fast pyrolysis process and unfortunately some remaining solubility after the pyrolysis. In order to investigate this material in a transistor, first a homogenous thin film after pyrolysis needs to be obtained. The material was dissolved in chlorobenzene (20 mg per ml) and spin coated on substrates cleaned as described above. All these steps were performed in the controlled nitrogen atmosphere within a glovebox. The resulting film had a layer thickness of around 100 nm (measured with ellipsometry). Following, the sample was heated up in the heating stage according to the in Section 3.2.2.2 described recipe. The results of TGA measurements were used as start parameters for the pyrolysis process.

It was the first investigated polymer thin film and it showed no differences in the results no matter which temperature ramp was used, only the end temperature was important. After the temperature reached 230 °C no further changes in the layer could be observed. On top of Figure 6.9, a photo of two substrates is shown. The substrate on the left is the one before pyrolysis process and the one on the right is the substrate after pyrolysis process. A clear change in the color can be observed, which is probably assigned to

---

[3] Type of size exclusion chromatography which separates analytes on the basis of size.

6. Organic Semiconductors with thermally activated solubility reduction

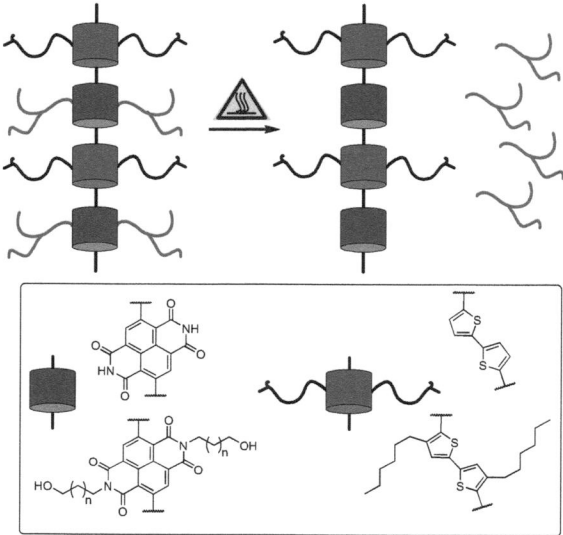

Figure 6.7: Concept of the thermally switchable solubility reduction on polymers - NDI:bitiophene copolymers.

the interference effect caused by the decrease of the layer thickness (around 30%) and not by the change in the band gap. The layer thickness decrease corresponds very well to the calculated weight loss value of 26.3%. Optically, no inhomogeneities were visible on the pyrolyzed sample.

6.3.1.1. Morphology

By performing some AFM measurements on the samples after the pyrolysis process, changes in the morphology of the thin layer can be found. On the bottom of the Figure 6.9, the AFM picture and its 3D representation after pyrolysis is shown. Although the layer seems to be still homogeneous even after the pyrolysis, a lot of holes with diameters above 200 nm are measured. The holes in the layer can probably be explained by the volume contraction. Another possible explanation, as figured out later, could be that the used layer thickness was too high, and prevented the evaporating side chains from leaving

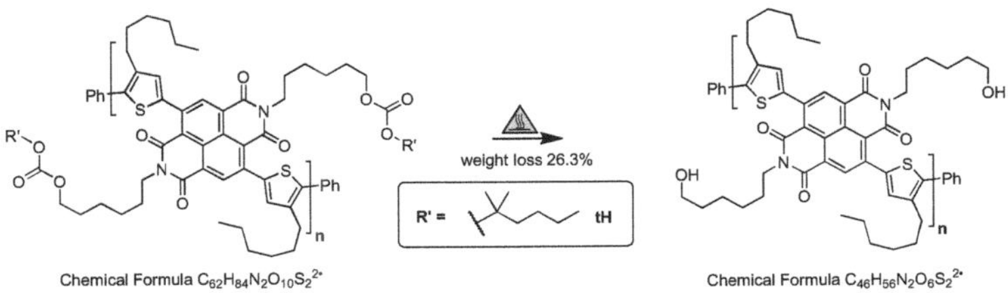

6.3. NDI Derivatives - Investigation of Synthesized Polymers

Figure 6.8: Chemical structure of P(HtHC-NDI-4HT2) and the corresponding result after the pyrolysis process.

the thin film without such drastic changes in the layer. The root-mean-square (RMS) roughness of the layer increased from 1.4 nm to 2.8 nm. The solubility test was negative - as can be seen at the corner of the pyrolyzed substrate the thin layer can be easily washed out using chlorobenzene solvent.

6.3.1.2. OFETs

With such homogeneous layers the first attempt on processing an OFET device was started. All OFET devices were prepared by cooperation partner Milan Alt (KIT) at InnovationLab in Heidelberg. The device layout used for this purpose can be looked up in Section 3.3.2.2. One has to mention that in this case no interfacial layer was used and 300 nm of parylene-C served as dielectric. Additionally, the parylene-C layer acts as encapsulation method against oxygen and humidity. Parylene-C dielectric was prepared on top of the active layer via vapor phase deposition from a PDS2010 Coating System by SPSTM. Characterization was carried out under ambient conditions in a 3 probe setup using an Agilent 4155C Semiconductor Parameter Analyzer. Transfer characteristics were measured in saturated regime. The data evaluation was performed by Milan Alt, a detailed description of these methods is not discussed within this work. Further information about the OFET data evaluation methods can be found in [146, 207]. Figure 6.10 shows transfer characteristics of OFET devices before and after pyrolysis. All samples show transistor behavior. The sample after only one minute pyrolysis process (red line) provides highest mobility and the best transfer characteristic. It is not clear if the pyrolysis process is completely done at this point. The samples after

6. Organic Semiconductors with thermally activated solubility reduction

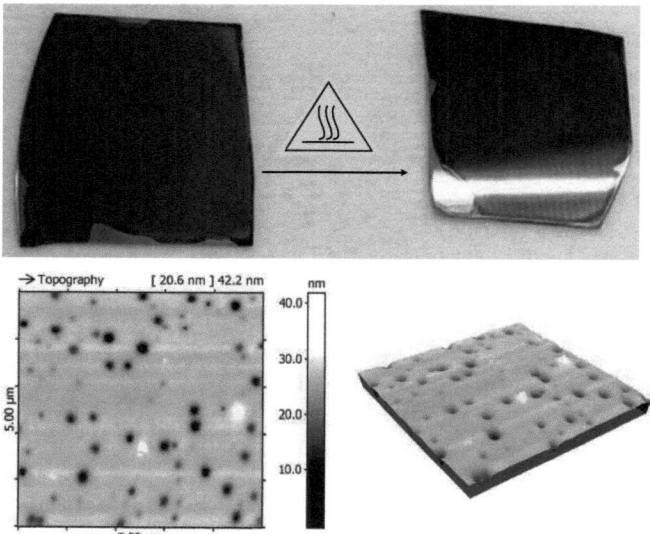

Figure 6.9: Effect of the pyrolysis process of P(HtHC-NDI-4HT2) thin film on a silicon substrate (top). Strong change in the color can be observed, which probably is related to the interference effect caused by the decrease of the layer thickness (around 30%). On the bottom an AFM picture and its 3D representation after pyrolysis is presented. The layer seems to be still homogenous after the pyrolysis, however due to the volume contraction many holes appear. The root-mean-square roughness is around 2.8 nm.

5 and 25 minutes of thermal treatment, where the pyrolysis process is done, show a significant decrease in the performance (green and blue line), however they are still better than the untreated substrate (black line). The highest achieved mobility is $\mu = 2.0 \cdot 10^{-4} \mathrm{cm}^2/\mathrm{Vs}$ with an on/off ratio of 10^2. The high performance of the hero device is probably caused by some morphological changes (not completed pyrolysis process) and should be not over-interpreted. Unfortunately, only devices with a channel length of 50 µm were working so no TLM could be used to derive the contact and channel resistance. No further devices were prepared and no further analytic measurements like PES or IR were performed on this polymer due to insufficient amount of available

6.3. NDI Derivatives - Investigation of Synthesized Polymers

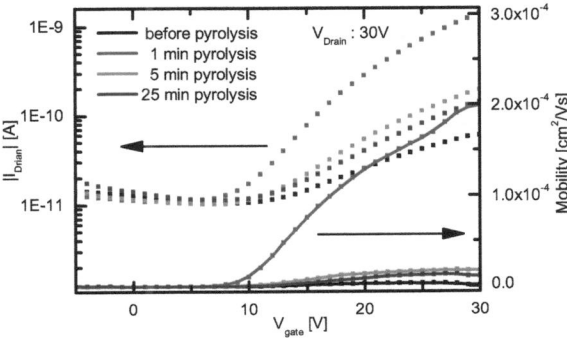

Figure 6.10: OFETs transfer characteristics before (black) and after (red, green, blue) pyrolysis process. Different pyrolysis periods were tested. The applied V_{SD} is 30 V.

material.

6.3.2. P(HtHC-NDI-4HT2) → P(NDI-C6OH-4HT2) Molecular Weight of 200 kDa

The investigated P(HtHC-NDI-4HT2) polymer with molecular weight of 200 kDa provides results very close to the requirements described at the beginning of this chapter. The samples were prepared in the same manner as described above but with lower concentration (10 mg per ml). The layer thickness of the prepared samples was around $\sim$ 38 nm. As expected after the pyrolysis the thickness decreased to 24 nm which corresponds to a volume contraction of 36%, which again is in a good agreement with the calculated value. Optically almost no difference between sample before and after pyrolysis could be observed, only a small change in the color hue was found.

6.3.2.1. Morphology

In order to investigate the surface of the prepared samples, AFM measurements were performed. Figure 6.11 shows the corresponding results. On the left side

6. Organic Semiconductors with thermally activated solubility reduction

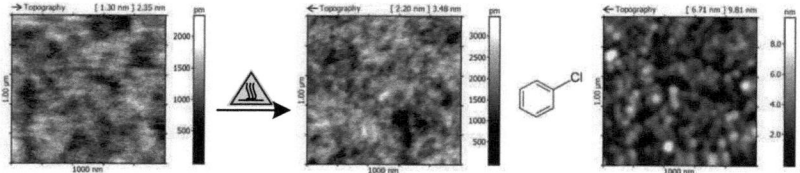

Figure 6.11: AFM images of P(HtHC-NDI-4HT2) before the pyrolysis, after the pyrolysis and after solubility test with chlorobenzene solvent which was used to dissolve the origin material. An increase of RMS roughness can be observed from 0.26 nm before pyrolysis, 0.39 nm after pyrolysis to 1.20 nm after 60 s immersion in chlorobenzene.

the AFM image of the untreated spin coated layer is depicted. It shows a very smooth surface with a calculated RMS value of 0.26 nm which increases after the pyrolysis to 0.39 nm. Again, the pyrolyzed substrate was immersed in chlorobenzene for one minute. The AFM measurement of this sample shows significant increase in the roughness of the thin film to 1.20 nm. In Figure 6.12 the influence of the chlorobenzene solvent on the thin film is shown. In case of the not pyrolyzed sample (left side) the polymer layer is completely dissolved and washed away. On the sample after the pyrolysis optically almost no change was observed. Only a thin "coffee stain" can be barely seen (red arrow marks the "coffee stain"). It shows the border where the solvent droplet ended, before it was dried out with nitrogen. An AFM measurement was performed exactly at this line. The result is presented on the rightmost side of the Figure 6.12. It is a 3D representation of the AFM image and the edge is five times thicker ($\sim$ 200 nm) than the layer itself. This effect and the previously described increase of the surface roughness validate the assumption that the thin film even after the pyrolysis process will still have a residual solubility. To confirm this impression PES and IR measurements were performed.

6.3.2.2. XPS and IR results

The PES measurements were performed in collaboration with TU Darmstadt by Marc Hänsel, a former master student. All samples used for PES and IR measurements were prepared in the same manner (described in Section 6.3.1).

6.3. NDI Derivatives - Investigation of Synthesized Polymers

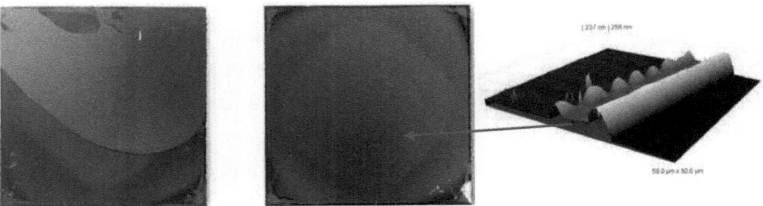

Figure 6.12: Photos of two prepared samples of P(HtHC-NDI-4HT2) treated with chlorobenzene solvent before (left) and after (middle) pyrolysis process. The red arrow shows the optically visible difference between surface treated with chlorobenzene and the untreated one. The image on the right side shows a 3D representation of an AFM measurement performed directly on the visible "coffee stain".

After the preparation process the samples were immediately transferred to the respective measurement technique. In Figure 6.13, the measured XP spectra are shown. At the top of this figure, the survey spectra are depicted, which indicate which elements are in the used polymer. All expected elements were assigned. Only one additional element was found, namely silicon. In this case it is the substrate which can be seen through the spin coated thin film. The characteristic SiO_2 doublet is shown in the Si2p core level spectrum, however only in the pyrolyzed sample, which again confirms the decrease in layer thickness. The components of S2p and N1s core levels remain almost unchanged after the pyrolysis process. The small difference in signal intensity can be explained by the usage of two different samples. In order to avoid any contamination from clean room atmosphere, two samples were prepared in a nitrogen filled glovebox: one before and one after pyrolysis process. The pyrolysis process was performed in a glovebox as well so that both samples were mounted at the same PES sample holder, measured at the same time with the same period of time in clean room atmosphere. Therefore, very small variations in layer thickness and with that signal intensity are conceivable. In the C1s and O1s detail spectra (bottom) the expected change in the components is shown. In order to understand and describe the effects in detail, fitting of these core level measurements was performed. Figure 6.14 shows the results of this procedure. At the top of this figure, the O1s detail spectra before (left) and after (right) pyrolysis are shown. Again the fit and

6. Organic Semiconductors with thermally activated solubility reduction

terminology is similar to that proposed in [154]. Black dots represent the measured spectrum with removed background and the red line shows the resulting fit. At the bottom of each graph an error curve is plotted, which describes the deviation between the fit and measurement (the error curve is 3 times magnified). In case of the O1s measurement before pyrolysis process the fit was done employing three singlets in order to assign three oxygen species, which are present within the investigated polymer. The assignment can be followed by the corresponding colors within the chemical structure. The small inconsistency in ratio between the singlets occur probably due to some defects in the polymer itself. After pyrolysis process, the carbonates should be gone as depicted in Figure 6.8 and an OH group should appear. Such a behavior is shown in the O1s core level spectra after pyrolysis (top-right). In this case, as expected, only two singlets were needed to fit the spectra. However, the green highlighted singlet appears to have too much intensity in comparison to the blue one. This could be the next confirmation that some residual side chains are still present on the surface. The C1s core level spectrum shows a complex component where five different carbon species can be found. The peaks marked with green and cyan color are not assigned within the chemical structure. An exact assignment of these peaks brings no advantages and also exceeds the capabilities of the measuring system. Two carbons which are bound to three oxygen atoms have the highest oxidation state within this polymer and as expected the singlet corresponding to these carbon atoms is chemically shifted to the higher binding energies. The same applies for the other species. After the pyrolysis process two carbon species are affected. The species marked with orange color disappear and the one marked with dark blue color decreases in intensity as expected. In this case, no residues from the side chains are detected. UPS measurements were also performed, however they are not discussed within this work.

All IR measurements presented in this study were performed in collaboration with University of Heidelberg by Sabina Hillebrandt, former master student. In Figure 6.15, a comparison of IR measurements between bulk spectra (bottom) and thin film (top) before (black) and after (red) pyrolysis is shown. In the bulk spectra, the absorption corresponding to the vibrations of the carbonates disappears completely after the pyrolysis process (marked by green rectangle

6.3. NDI Derivatives - Investigation of Synthesized Polymers

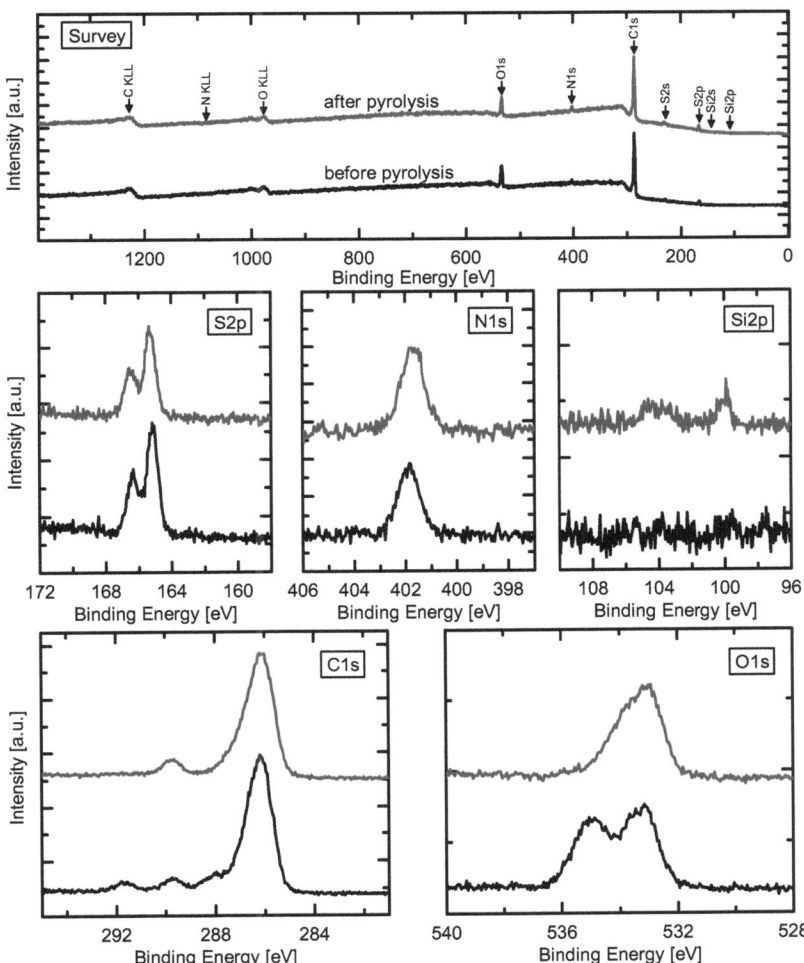

Figure 6.13: P(HtHC-NDI-4HT2) XP survey spectrum (top) before and after pyrolysis and detail spectra of all elements which can be found in the used polymer (middle and bottom rows). The Si2p emission line which belongs to the underlying substrate and is not part of the polymer, can only be seen in the pyrolyzed layer, which confirms the decrease of the layer thickness.

6. *Organic Semiconductors with thermally activated solubility reduction*

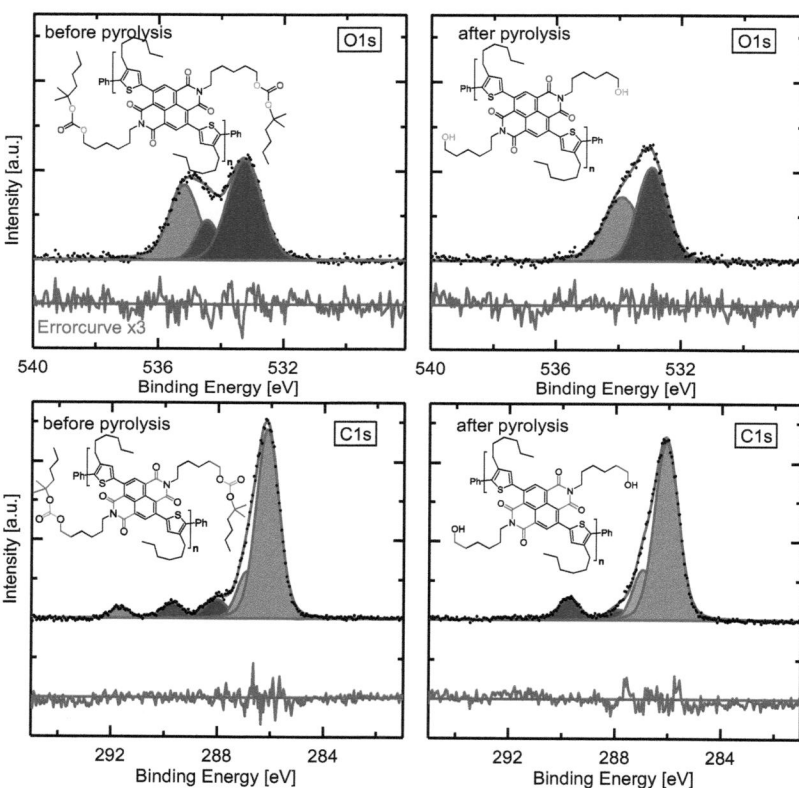

Figure 6.14: The O1s and C1s core level spectra (black dots represents the measured spectrum, however with removed background) and the corresponding fits for polymer before (left side) and after (right side) pyrolysis. The colors of the fitted component are matched with the colors of the chemical structure, except for the two highest components in the C1s emission line which represent the rest of not assigned carbon atoms.

6.3. NDI Derivatives - Investigation of Synthesized Polymers

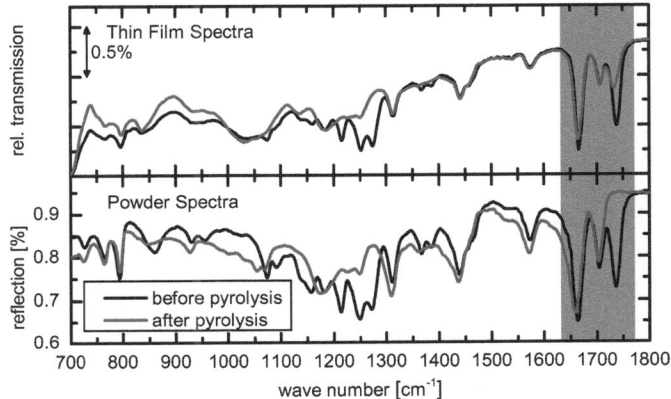

Figure 6.15: Comparison of IR spectra of powder (bottom) and a thin film (top) spin coated on a silicon substrate before (black line) and after (red line) pyrolysis. The green highlighted region indicates absorptions corresponding to the vibrations of the carbonates.

inside the graph). In case of the spin coated thin film, some remaining carbonates within the layer can still be detected. This corresponds also to the above presented PES results. One can assume that the cleavage of the side chains takes place but the decomposition to alcohol and CO_2 is kinetically inhibited. The chlorobenzene treatment after pyrolysis process confirms this idea. In Figure 6.16, the spectrum after solvent treatment (blue) shows no absorptions corresponding to the vibrations of the carbonates. This can be explained by the decrease in viscosity of the polymer film, which allows to overcome the kinetic inhibition and the pyrolysis process can be completed by releasing the CO_2. The strong change in the base line is probably caused by some inhomogeneities. In the same figure the absorptions corresponding to the CH-stretch vibrations are shown, their change in the intensity corresponds very well to the calculated value (weight loss - see Table 6.2).

6. Organic Semiconductors with thermally activated solubility reduction

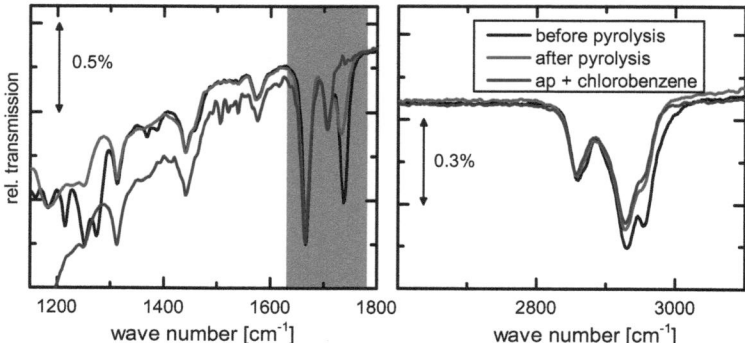

Figure 6.16: IR spectra of the spin coated polymer before (black line) and after (red line) pyrolysis process (same as in Figure 6.15). Additionally, a measurement after chlorobenzene treatment is shown (blue line). The green highlighted region indicates absorptions corresponding to the vibrations of the carbonates. On the right side the absorptions corresponding to the CH-stretch vibrations can be found.

6.3.2.3. OFETs

In order to investigate the electrical properties of P(HtHC-NDI-4HT2) before and after pyrolysis process, some OFET devices were prepared and evaluated by PhD student Milan Alt from KIT. The device layout used for this purpose is the same as in previous section. One has to mention that in this case an interfacial layer was used, namely Juls SAM (see Section 5.4) in order to improve the electron injection in the semiconductor layer. Transfer characteristics were measured in the linear regime. In Figure 6.17 the OFET devices before and after pyrolysis are shown and both show good transistor behavior. All data shown in this study are average values from at least 5 devices. The transistors after pyrolysis process (red line) show higher mobility ($\mu = 5.0 \cdot 10^{-5}$ cm^2/Vs vs. $\mu = 1.0 \cdot 10^{-6}$ cm^2/Vs) and an on/off ratio of 10^4. Devices with various channel length were measured, which allow to perform the TLM data evaluation and to derive the contact and channel resistance (Figure 6.17 right side). The red curve represents the samples after pyrolysis process and it shows lower channel resistance. Obviously even the remaining side chains within the layer

6.3. NDI Derivatives - Investigation of Synthesized Polymers

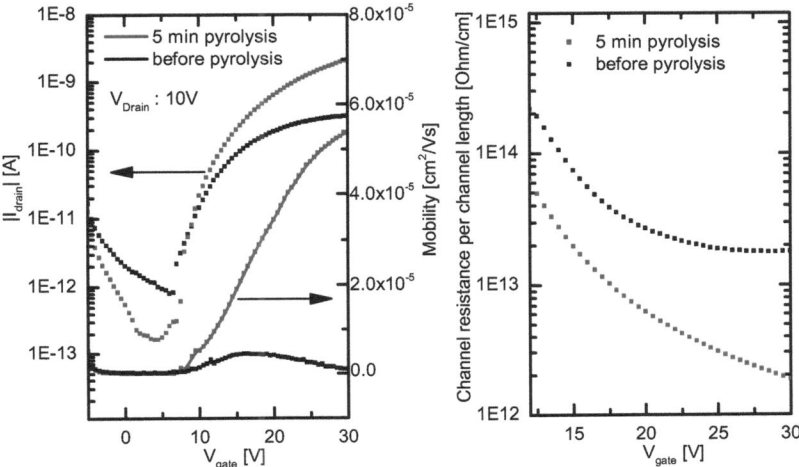

Figure 6.17: Left: OFET transfer characteristics before (black) and after (red) pyrolysis process. The applied V_{SD} is 10 V. Right: channel resistance per channel length derived from TLM.

(see IR measurements in the previous section) are not hindering the electrical properties of the investigated devices. The channel resistance could probably be even lower if the pyrolysis process would be 100% successful. More detailed analysis of the OFET devices will be found in the planned dissertation of Milan Alt.

6.3.3. P(tHC-NDI-4HT2) → P(NDI-4HT2)

The next investigated polymer was P(tHC-NDI-4HT2), which due to the used NDI-carbamates should eliminate the remaining solubility after the pyrolysis process observed on the previously presented compound. The chemical structure of this polymer and the concept of the pyrolysis process is shown in Figure 6.18. After the thermal decomposition of the precursor polymer the P(NDI-4HT2) polymer is obtained with free imide functions which are able to act as intermolecular linking sites and reduce the residual solubility of the polymer. The TGA measurements and GPC analysis performed by

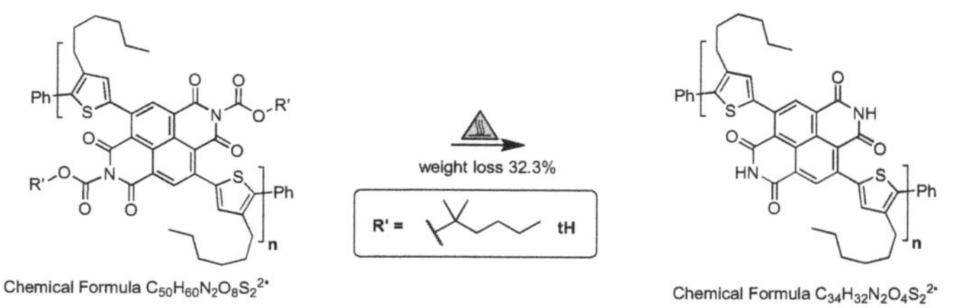

6. Organic Semiconductors with thermally activated solubility reduction

Figure 6.18: Chemical structure of P(tHC-NDI-4HT2) and result after the pyrolysis process.

Torben Adermann (OCI) promise both a much lower pyrolysis temperature (see Table 6.2) and that after the pyrolysis the resulting material is insoluble. The synthesis and full description of this polymer can be found in [184]. The investigated samples were prepared in the same manner as described in previous section. The layer thickness of the prepared samples was around ~ 33.5 nm (measured by ellipsometry). As expected after the pyrolysis the thickness decreased to 27 nm which corresponds to a volume contraction of 20%, which compared with the expected weight loss of 30% is rather low. Optically almost no differences between the pre and post pyrolysis samples could be observed, only a small change in the color hue was found. The same applies to the sample after one minute chlorobenzene solvent treatment. Moreover, no change in the layer thickness was measured.

6.3.3.1. Morphology

The AFM investigations of the prepared samples show a very homogenous surface (see Figure 6.19). On the left side, the AFM image of the untreated spin coated layer is presented. It shows a very smooth surface with a RMS roughness value of 0.23 nm. In the middle, the sample after the pyrolysis is presented. It shows small changes of the surface morphology, however no change in the roughness itself was measured. The pyrolyzed substrate was immersed in chlorobenzene for one minute and also measured via AFM. The sample shows minor increase in the roughness of the thin film (0.31 nm). These results confirm the expectations that the polymer films obtained after the pyrolysis are sufficiently resistant to solvent treatment.

6.3. NDI Derivatives - Investigation of Synthesized Polymers

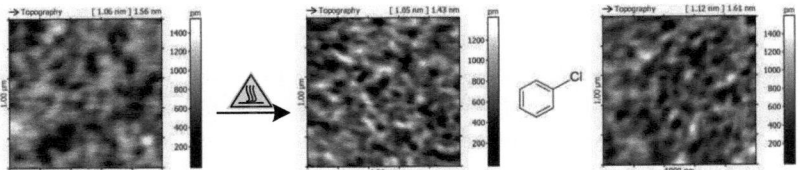

Figure 6.19: AFM images of P(tHC-NDI-4HT2) before the pyrolysis, after the pyrolysis and after solubility test with chlorobenzene which was used to dissolve and process the initial material. Almost no changes in RMS roughness can be observed: 0.23 nm before pyrolysis, 0.24 nm after pyrolysis and only 0.31 nm after 60 s immersion in chlorobenzene.

6.3.3.2. XPS and IR

The same set of samples was also investigated with IR and XPS measurements. In Section 6.3.2.2 all measured XPS spectra from P(HtHC-NDI-4HT2) polymer were presented in order to show and describe the measurement procedure. Here only the relevant spectra with removed background and with the corresponding fits are shown (see Figure 6.20). For unprocessed data and further measured spectra see Chapter A, Figure A.5. The PES measurements were performed by Marc Hänsel (TU Darmstadt). All samples used for PES and IR measurements were prepared in the same way (already described in Section 6.3.1) and immediately transferred to the respective measurement technique. In Figure 6.20 fits of O1s (top) and C1s (bottom) core levels before (left) and after (right) pyrolysis are presented. The fit and terminology is the same as described in Section 6.3.2.2. Three singlets were employed to assign the three different oxygen species within the investigated polymer. The assignment can be followed by the corresponding colors within the chemical structure. Some defects in the polymer itself are responsible for the visible differences in ratio between the violet and green color marked singlets. After the pyrolysis the carbamates groups should be gone as depicted in Figure 6.18 and a free imide function should remain. The resulting measurements are presented in the O1s detail spectra after pyrolysis (top-right). As expected, the two singlets (violet and green marked components) almost completely disappeared. Only a very small shoulder is still present which can be assigned to some

6. Organic Semiconductors with thermally activated solubility reduction

residuals of the side chains or maybe to some very small contamination from the clean room atmosphere. The complex emission lines of the C1s core level show five different components, which indicates five carbon species within the investigated polymer. In the chemical structure (colored elements) all carbon species are marked. Two carbons which are bound to two oxygen atoms have the highest oxidation state within this polymer and as expected the singlet corresponding to these carbon atoms is chemically shifted to the higher binding energies. The same applies for the other species: four carbon-oxygen double bonds (violet marked component) and carbon bound to one oxygen (dark blue component). As expected after the pyrolysis process, two carbon species are gone. The species marked with orange and dark blue color disappeared completely and the ratio between the two main components and the violet marked one changed as expected. In this case no residues from the side chains are detected. UPS measurements were also performed, however they are not going to be discussed within this work.

The IR measurements were performed in collaboration with University of Heidelberg by Sabina Hillebrandt. In Figure 6.21 a comparison between IR spectra from bulk material (bottom) and thin film (top), both before (black line) and after (red line) pyrolysis at 180 °C is given. Bulk spectra as well as thin film spectra show a very similar behavior. In both cases, the absorption corresponding to the vibrations of the carbamates disappears completely after the pyrolysis process (region marked by green rectangle inside the graph). This corresponds very well to the above presented PES results. Moreover, the two isolated carbonyl vibrations of the NDI-carbamates (1684 cm^{-1} and 1712 cm^{-1}) merge after the pyrolysis into a single broad peak (1692 cm^{-1}, red), which is caused due to carbamates transformation into free imide within the pyrolysis process (see Figure 6.18). In Figure 6.22 the corresponding N-H vibrations are also presented (red and blue line around 3194 cm^{-1} and 3075 cm^{-1}). The very weak absorptions which were observed in the sample prior the pyrolysis process (black line) indicate that a minor carbamates decomposition also takes place before the pyrolysis. This observation is consistent with the results of TGA analysis performed by Torben Adermann (OCI). More important is however the observation that the IR measurements performed on the sample after pyrolysis and after cholorobenzene treatment remain unchanged (compare blue

6.3. NDI Derivatives - Investigation of Synthesized Polymers

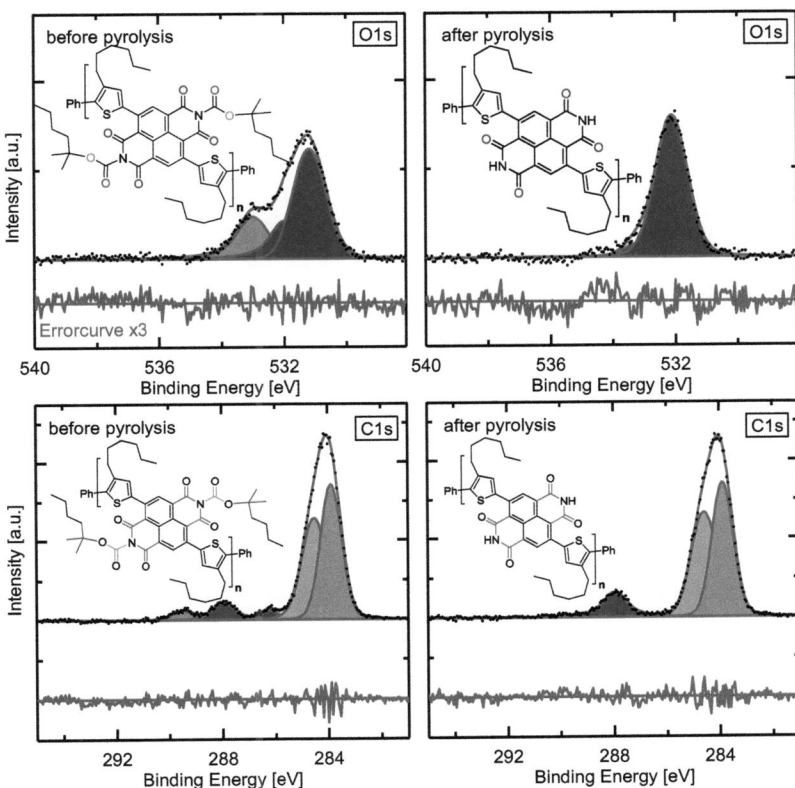

Figure 6.20: The O1s and C1s core level spectra (black dots represents the measured spectrum, however with removed background) and the corresponding fits for polymer before (left) and after (right) pyrolysis. The colors of the fitted component are matched with the colors of the chemical structure, except for the two highest components in the C1s emission line which represent the rest of not assigned carbon atoms. The small component in the O1s line (violet) after pyrolysis indicates probably some very small residuals from thermally labile side chains or some small contamination. An exact assignment is not possible in this case.

6. Organic Semiconductors with thermally activated solubility reduction

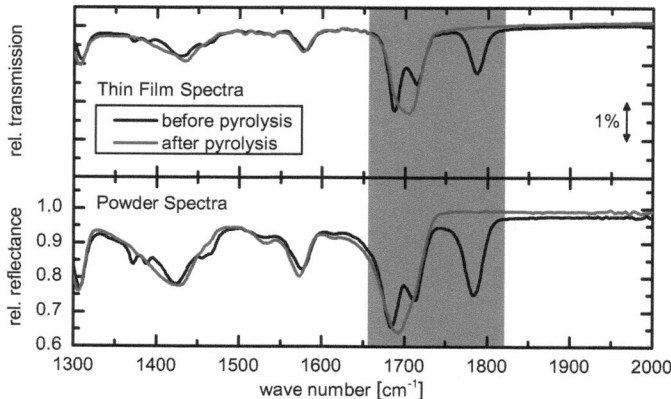

Figure 6.21: Comparison of IR spectra of powder (bottom) and a thin film (top) spin coated on a silicon substrate before (black line) and after (red line) pyrolysis. The green highlighted region indicates absorptions corresponding to the vibrations of the carbamates.

and red line). This confirms again the previous statement that the pyrolyzed thin film of P(tHC-NDI-4HT2) shows a very good resistance to solvents.

6.3.3.3. OFETs

In the next step OFET devices were prepared in order to investigate the electrical properties of the P(tHC-NDI-4HT2) before and after pyrolysis process. The devices were prepared and evaluated by collaboration partner Milan Alt (KIT). The same device layout was used as in the previous section. Also in this case the same interfacial layer was used in order to improve the electron injection in the semiconductor layer. Transfer characteristics were measured in the linear regime. On the left of Figure 6.23, transfer characteristics from OFET devices before and after pyrolysis are presented. All data shown in this study are average values from at least 5 single devices. All OFET devices show a constant on/off ratio of 10^3. The transistors after pyrolysis process (blue curve) show a much higher mobility of about $\mu = 4.0 \cdot 10^{-5}\,\text{cm}^2/\text{Vs}$. The pyrolysis with higher temperature leads to even

6.3. NDI Derivatives - Investigation of Synthesized Polymers

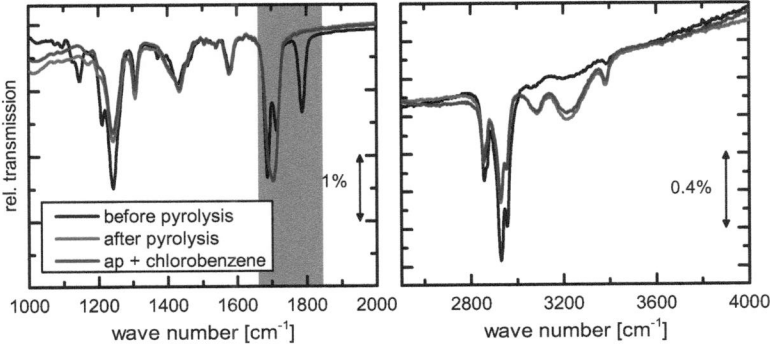

Figure 6.22: IR spectra of the spin coated polymer before (black line) and after (red line) pyrolysis process. Additionally, a measurement after chlorobenzene treatment is shown (blue line). The green highlighted region indicates absorptions corresponding to the vibrations of the carbamates. On the right side the absorptions corresponding to the CH-stretch and N-H vibrations are presented.

slightly better performance ($\mu = 4.8 \cdot 10^{-5}$ cm^2/Vs) and shows a mobility an order of magnitude higher than the device before the pyrolysis process. Devices with various channel length were measured which allows to perform the TLM data evaluation and to derive the contact and channel resistance (Figure 6.23 right side). Obviously, the devices after pyrolysis show a much lower channel resistance. The effect is stronger than in case of the polymer presented in the previous section which maybe indicates an influence of carbamates on electrical properties. In case of P(HtHC-NDI-4HT2) polymer, after the pyrolysis, some OH groups remain on the polymer backbone and maybe act as traps for the majority carriers, which are not present in the current case. No further measurements were performed on this polymer due to a limited amount of available material.

6.3.4. P(HtODC-NDI-T2) → P(NDI-C6OH-T2)

The P(HtODC-NDI-T2) polymer presented in Figure 6.24 has no alkyl substituents on the tiophene units. In order to maintain the solubility significantly larger alkyl groups were added to the thermally labile carbonate functions. It

6. Organic Semiconductors with thermally activated solubility reduction

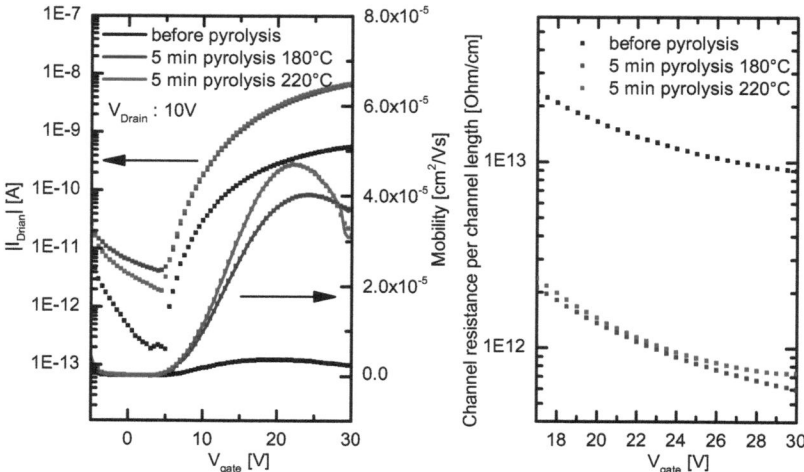

Figure 6.23: OFET transfer characteristics before (black) and after (blue, red) pyrolysis process (left). The applied V_{SD} is 10 V. On the right side the channel resistance per channel length derived by usage of TLM is presented.

should lead to two effects: it should improve the resistance to solvents after the pyrolysis, however a large amount of low volatile alkenes arises in the layer and a very high weight loss is expected, which could lead to defects in the spin coated layer. The TGA and GPC measurements performed by Torben Adermann (OCI) show a two step pyrolysis (for more details see [184]) and promise, as expected, a lower solubility after the pyrolysis process than for the P(HtHC-NDI-4HT2) polymer. In order to investigate the polymer and its two step pyrolysis, following samples were prepared:

- before pyrolysis - sample without thermal treatment

- 3 minutes pyrolysis at 200 °C - carbonates pyrolysis without evaporation of alkene products

- 3 minutes pyrolysis at 220 °C - complete carbonates pyrolysis with the onset of evaporation of alkene products

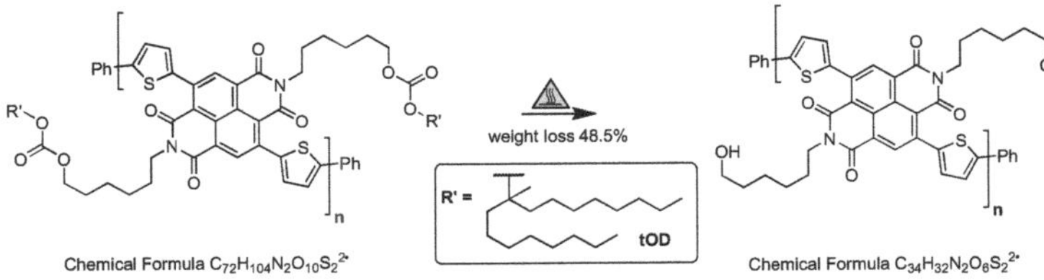

6.3. NDI Derivatives - Investigation of Synthesized Polymers

Figure 6.24: Chemical structure of P(HtODC-NDI-T2) and the corresponding result after the pyrolysis process.

- 20 minutes pyrolysis at 220 °C - complete carbonates pyrolysis with slow but complete evaporation of alkene products
- 3 minutes pyrolysis at 250 °C - complete carbonates pyrolysis with fast and complete evaporation of alkene products - only for IR and AFM measurements due to limited amount of available material.

All investigated samples were prepared in the same manner as described in the previous section. The layer thickness of the prepared samples was around ~ 39 nm. Optically almost no difference between samples before and after pyrolysis could be observed, only a small change in the color hue is visible. The same applies to the sample after one minute chlorobenzene solvent treatment moreover no change in the layer thickness can be measured.

6.3.4.1. Morphology

All prepared samples were measured via AFM and ellipsometry, a summary of the results is presented in the Table 6.3. All samples after chlorobenzene treatment show a very similar layer thickness (16.1 nm → 17.0 nm), whereas samples only after pyrolysis process vary between 17.4 nm (20 minutes at 220 °C) and 31.5 nm (3 minutes at 200 °C). The cleavage of the thermally labile groups seems to be complete for all samples but the remaining alkene groups are first removed from the layer after solvent treatment. This can be clearly seen on the sample which was pyrolyzed only 3 minutes at 200 °C and the sample pyrolyzed 20 minutes at 220 °C. The first sample has a layer thickness of 31.5 nm after the pyrolysis, whereas the other one has a layer thickness of

6. Organic Semiconductors with thermally activated solubility reduction

Pyrolysis Time		RT w/o CB	200°C w/o CB	200°C w/ CB	220°C w/o CB	220°C w/ CB	250°C w/ CB
0 min	Thick.	39 nm	-	-	-	-	-
	RMS	0.39 nm	-	-	-	-	-
3 min	Thick.	-	31.5 nm	16.3 nm	27.8 nm	16.1 nm	17.0 nm
	RMS	-	0.69 nm	0.84 nm	0.47 nm	0.51 nm	0.48 nm
20 min	Thick.	-	-	-	17.4 nm	16.9 nm	-
	RMS	-	-	-	0.37 nm	0.44 nm	-

Table 6.3: Ellipsometry (layer thickness determination) and AFM (roughness calculations) results of thin film measurements before and after pyrolysis process as well as after additional chlorobenzene (CB) solvent treatment. Empty cells indicate no measurement for the corresponding sample.

only 17.4 nm. After the solvent treatment, both samples have a very similar layer thickness of 16.3 nm and 16.9 nm, respectively. Therefore, an assumption is possible that in the first sample, directly after the pyrolysis process, a much larger amount of remaining alkene groups is still contained in the thin film. A representative selection of AFM measurements is presented in Figure 6.25. One can observe a trend in the RMS roughness calculations. Samples exposed to a longer thermal treatment show a smoother surface and after the solvent treatment the roughness increases only marginally. There are no additional defects like holes or cracks in the layer even though in these samples almost all alkene products are evaporated and the highest volume contraction occurs. This behavior confirms also the increase of solvent resistance. Overall, one can observe that with a higher thermal load, smoother surface and higher solvent resistance of the layer is achieved.

6.3.4.2. XPS and IR

IR and PES measurement were performed on the same set of prepared samples. All samples were prepared in the same manner (already described in Section 6.3.1). After the preparation process in the glovebox, the samples were immediately transferred through ambient clean room conditions to the respective measurement technique. In Figure 6.26 the measured XP spectra are shown. All PES spectra presented in this section were measured by Lena

6.3. NDI Derivatives - Investigation of Synthesized Polymers

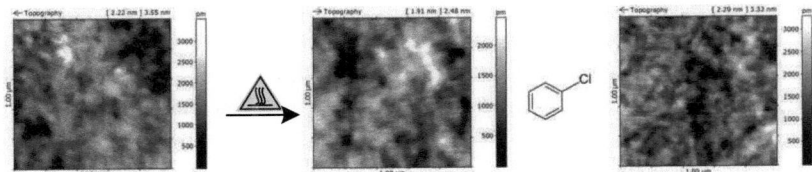

Figure 6.25: Representative AFM images of P(HtODC-NDI-T2) before the pyrolysis, after the pyrolysis (20 minutes at 220 °C) and after solubility test with chlorobenzene which was used to dissolve the starting material. No significant change in RMS roughness can be observed: 0.39 nm before pyrolysis, 0.37 nm after pyrolysis and 0.44 nm after 60 s immersion in chlorobenzene. The remaining AFM measurements are presented in Appendix A Figure A.7.

Kuske (TUD), a former bachelor student. Only C1s (left) and O1s (right) core level spectra are shown and discussed within this section. For all additional measurements see Chapter A, Figure A.6. In Figure 6.20, fits of the above mentioned core levels before (bottom) and after the pyrolysis process are presented. Black dots represent the measured spectrum with removed background, the red line shows the resulting fit. To fit all the components from the O1s emission line, three singlets were employed to assign the three different oxygen species within the investigated polymer. The assignments are matched to corresponding colors in the chemical structure. At the bottom the XP spectra of a non-pyrolyzed sample is presented. The blue and green marked components have, as expected, almost the same intensity. The third component (violet) has only half the intensity, which corresponds very well to the chemical structure with three different oxygen species. The sample after 3 minutes pyrolysis at 200 °C still show three components, however with a different ratio. The component which belongs to the NDI core is the most pronounced one. The other two components decrease in intensity, which indicates the start of the pyrolysis process. The next two samples show a similar behavior, however the middle component (marked with green for this and the next sample) increases in intensity. It shows probably a mix of the residuals of the thermally labile groups as well as the first OH groups, which should form during the pyrolysis process. The same behavior shows the sample pyrolyzed for 20 minutes at 220 °C. A similar picture provide the C1s core

6. Organic Semiconductors with thermally activated solubility reduction

level measurements. Before the pyrolysis process, five different components are fitted. The corresponding assignment is enclosed in the chemical structure (colored elements). Two carbons which are bound to three oxygen atoms have the highest oxidation state within this polymer and as expected the singlet corresponding to these carbon atoms is chemically shifted to higher binding energies (orange marked feature). The same applies for the other species: four carbon-oxygen double bonds (violet marked component) and carbon bound to one oxygen (dark blue component). The first sample, similar to the O1s measurement, indicates only a start of the pyrolysis. A decrease in the orange marked component as well as some changes in ratio between the components are observed. In the next two steps, the peak marked with orange color disappears or most probably it is still there but is below the detection limit, and the ratio between the two main components and the components marked with violet and blue changes as expected. These XPS measurements confirm the presence of residual alkene groups and the incomplete pyrolysis process. UPS measurements were also performed, however they are not going to be discussed within this work.

All IR measurements shown in this section were performed in collaboration with University of Heidelberg by Sabina Hillerbrandt. In Figure 6.27, a comparison between IR spectra of a thin film before pyrolysis (black curve), after pyrolysis (various temperature and time, see Table 6.3) and after additional solvent treatment is presented (darker shade of each color). In case of samples pyrolyzed at temperature 220 °C and lower, a clear peak corresponding to the vibrations of the carbonates (1739 cm^{-1}) after pyrolysis is still present, which corresponds very well to the above presented XPS and AFM results. However, just like in the case of P(HtHC-NDI-4HT2) polymer, the residuals of the thermally labile groups disappear after additional 60 seconds of solvent treatment (darker shade of each color), which is very consistent with the ellipsometry measurements where each sample after the solvent treatment have a very similar layer thickness. The same behavior is shown on the right side of this figure, where CH-stretch vibrations are presented. In case of the sample pyrolyzed for 3 minutes at 250 °C, no residuals of the added side chains are detectable directly after pyrolysis process and there is no change in the spectra even after 60 seconds chlorobenzene treatment. The same

6.3. NDI Derivatives - Investigation of Synthesized Polymers

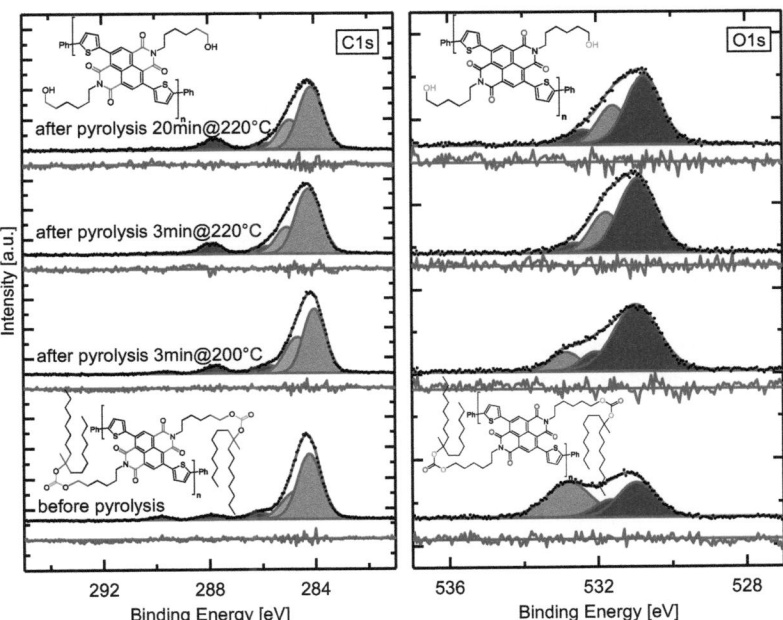

Figure 6.26: The O1s (right) and C1s (left) core level spectra (black dots represents the measured spectrum, however with removed background) and the corresponding fits of polymer before (bottom) and after (remaining spectra) three different pyrolysis processes introduced at the beginning of this section. The colors of the fitted component are matched with the colors of the chemical structure, except for the two highest components in the C1s emission line which represent the rest of not assigned carbon atoms.

6. Organic Semiconductors with thermally activated solubility reduction

applies to the CH-stretch vibrations. The small intensity variations occur by some thickness differences between the samples and the fact that the measurement after pyrolysis and after solvent treatment were in fact taken on the same sample, but not exactly at the same position (only manual adjusting is available). The wave number shift of more than $3\,\text{cm}^{-1}$ (peak at $1666\,\text{cm}^{-1}$) occurs probably due to a change in the molecular structure. During pyrolysis process the thermally labile groups disappear and therefore influence, with high probability, the vibrations of the adjacent imide groups.

6.3.4.3. OFETs

In Figure 6.28 the summary of OFET devices measurements is presented. The devices were prepared (in the same manner as described in Section 6.3.1.2) and evaluated in collaboration with Milan Alt (KIT). One has to mention that in this case no interfacial layer was used and silver electrodes instead of gold were employed. Additionally, samples pyrolyzed 20 minutes at $180\,°\text{C}$ were prepared. Transfer characteristics were measured in linear and saturated regime (for transfer characteristics data see Chapter A, Figure A.8). All data shown in this study are average values from at least 5 single devices, providing the standard deviation as statistical error. The highest achieved mobility was $\mu = 2.4 \cdot 10^{-4}\,\text{cm}^2/\text{Vs}$ for the device pyrolyzed for 3 minutes at $200\,°\text{C}$. An on/off ratio of 10^4 was measured for this hero device. Devices with various channel lengths were measured, which allow to perform the TLM data evaluation and to derive the contact and channel resistance (Figure 6.28). As expected the device with the lowest channel resistance achieves the best performance. Interestingly, samples exposed to higher thermal load show a higher channel resistance, which means that a lower amount of remaining alkene groups is not necessarily better for electrical properties. Devices pyrolyzed with temperature $T \geq 200\,°\text{C}$, except the one with 3 minutes pyrolysis time, show a higher channel resistance which also means lower performance in most cases. Obviously, the high temperature or the morphology changes induced through the high temperature have negative influence on electrical properties of the thin film. The presence of free hydroxyl groups as compared with the presence of carbonate functions in the precursor polymer seems to be more favorable for the electron-conducting properties. The red and green lines in Figure 6.28

6.3. NDI Derivatives - Investigation of Synthesized Polymers

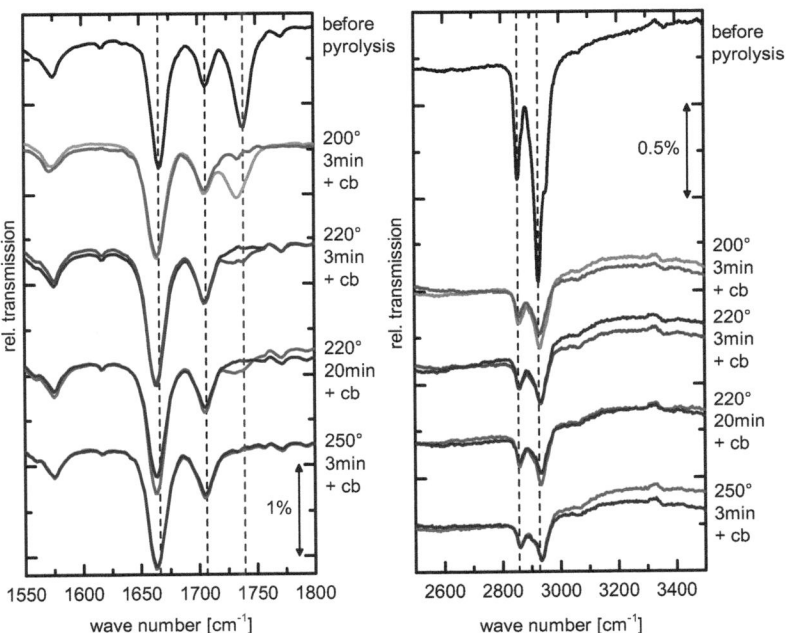

Figure 6.27: IR spectra of the spin coated polymer before (black line), after 3 minutes pyrolysis at 200 °C (green line), after 3 minutes pyrolysis at 220 °C (blue line), after 20 minutes pyrolysis at 220 °C (violet line) and after 3 minutes pyrolysis at 250 °C (red line). Additionally, a measurement after chlorobenzene treatment is shown (darker shade of each color). The region shown on the left side of this figure indicates absorptions corresponding to the vibrations of the carbonates. On the right side the absorptions corresponding to the CH-stretch vibrations are presented.

6. Organic Semiconductors with thermally activated solubility reduction

should only act as a guide to the eyes for the effect which again confirms the results from previous investigations. After pyrolysis with higher temperature or with longer period of time, the difference in the channel resistance between samples before and after solvent treatment decreases constantly. Probably the sample pyrolyzed at 250 °C, which shows no residuals of the added side chains in IR measurements would show no influence at all from the chlorobenzene solvent treatment. Unfortunately, no such device was prepared due to a limited amount of available material.

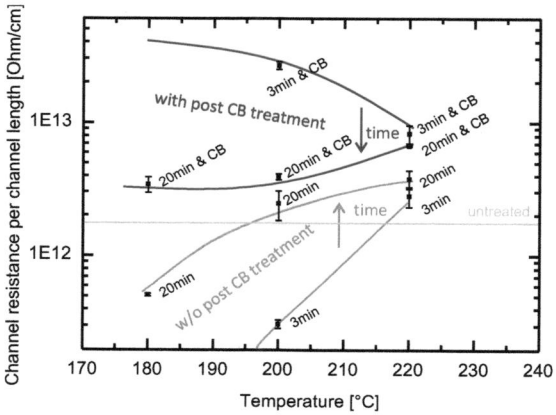

Figure 6.28: OFET characteristics of samples treated with various thermal load summarized to channel resistance per channel length derived by usage of TLM. The transfer characteristics in linear and saturated regime as well as the channel resistance plot for each device can be found in Appendix A, Figure A.8. The red and green curves should only act as a guide to the eyes.

6.3.5. Comparison and Discussion

Altogether, three different polymers (P(HtHC-NDI-4HT2) was available in two batches with different molecular weight) were investigated during this study. All of them were analyzed in detail using various analytical methods as well as in OFET devices. All polymers show very satisfactory wetting properties on various substrates, and no additional treatments were required

6.3. NDI Derivatives - Investigation of Synthesized Polymers

to achieve homogenous layers. The P(HtHC-NDI-4HT2) and P(HtODC-NDI-T2) polymers were synthesized according to the NDI-linker-carbonates approach (see Section 6.1). The pyrolyzed P(HtHC-NDI-4HT2) polymer still shows some residual solubility. However, through the pyrolysis process, better OFET performance is achieved. It seems that higher molecular weight and thinner layers help avoiding cracks and holes in the spin coated layer. For the P(HtODC-NDI-T2) polymer, a higher solvent resistance was presented. Moreover, it has been shown that a volume contraction of almost 50% does not lead to defects in the layer. However, device performances decrease after a complete pyrolysis process in comparison to the untreated or to only partially pyrolyzed devices, which can be ascribed to morphology modifications or even to the built-up stress in the layer. Interestingly, the residual alkene groups are not necessarily hindering electrical transport in the thin film. In order to achieve a complete carbonate pyrolysis, a temperature of 250 °C is needed, which in terms of flexible substrate usage may be problematic. Using this material, the highest mobility within the prepared OFET devices was achieved ($\mu = 2.4 \cdot 10^{-4}$ cm^2/Vs), however one has to mention that no stack optimizations to maximize these results were performed during this study. The P(tHC-NDI-4HT2) polymer with NDI-carbamates shows no residual solubility after pyrolysis process and much lower pyrolysis temperature. Only after 3 minutes of pyrolysis at 180 °C no remaining thermally labile side chains were detected within the layer. Moreover, the higher the pyrolysis temperature, the higher OFET performance could be achieved. However, according to [184] the synthesis process is very challenging and due to very low cleavage temperature of carbamates ($\sim$ 100 °C) the yield of the material can be very low. The highest mobility achieved with P(tHC-NDI-4HT2) was $\mu = 4.8 \cdot 10^{-5}$ cm^2/Vs which is an order of magnitude lower than in case of the P(HtODC-NDI-T2) polymer. All three materials deliver very smooth surfaces after pyrolysis. Even with additional solvent treatment after pyrolysis the measured RMS roughness in most cases does not exceed 1.0 nm, which is a satisfactory result.

6.4. Conclusion

In this work various new synthesized small molecules and polymers based on NDI were investigated. For small molecules, the goal of this project was unfortunately not reached. Indeed, several thin films were prepared from the small molecules, and even functioning OFET devices were prepared, however not after the pyrolysis process. Even though several dewetting problems during the pyrolysis process occurred (see Figure 6.29 first row) some interesting behavior could be observed, as for the PtHC-NDI, where the layer first changed to a melted mass and then if the given temperature is reached crystallized immediately. Moreover, most of the investigated small molecules have a pyrolysis temperature between 150 °C and 190 °C, which is the range where several flexible substrates can be used. Even though the many attempts with various NDI small molecules derivatives, none led to the desired results. It though still remains a promising method to achieve the goal of this project in near future.

A completely different picture is presented with the new synthesized and investigated polymers. All investigated polymers fulfilled most of the requirements. Representative summary of data is presented in Figure 6.29 (rows 2-4). The second row of this figure presents the AFM measurements after chlorobenzene treatment and the resulting RMS roughness. In case of P(HtHC-NDI-4HT2) and P(HtODC-NDI-T2) polymers some residuals of the thermally labile groups could be detected in the thin film even after pyrolysis process (see Figure 6.29, third row). This led to lower solvent resistance, which nevertheless should be sufficient enough for further preparation steps. Both negative and positive effects on electrical properties were observed due to the detected residuals after the pyrolysis (Figure 6.29, fourth row). However, a unique correlation could not be found. In case of P(HtODC-NDI-T2) polymer the OFET performance decreases after solvent treatment, probably the morphology changes induced through the immersion in chlorobenzene have negative influence on electrical properties of the thin film. Most probably an OFET device with thin film pyrolyzed at 250 °C would still show lower performance than before pyrolysis, however no change between the pre and post solvent treated sample would be observable. Unfortunately, no such device was prepared due to a limited amount of available material. Moreover, what can be said is that the high weight losses during the pyrolysis do not necessarily negatively affect the morphology, the same cannot be said about the electrical properties. Maybe the built-up stress in the layer is somehow hindering the transport properties. The P(tHC-NDI-4HT2) polymer was the only one which met all demands like processability, pyrolysis temperature and solvent resistance after pyrolysis. The only drawback is the achieved mobility in an OFET device, however again one need to say that no stack optimizations were considered in order to improve the device performance. Further optimizations in chemical structure are also possible, for further detail about the various possibilities see [184].

The achieved results are the first step to ease the process from solution where until now various problems still occur like intermixing of the used materials due to limited amount of orthogonal solvents. However, in order to provide competitive results to already established processes this approach needs to be transferred to some high potential materials.

6. Organic Semiconductors with thermally activated solubility reduction

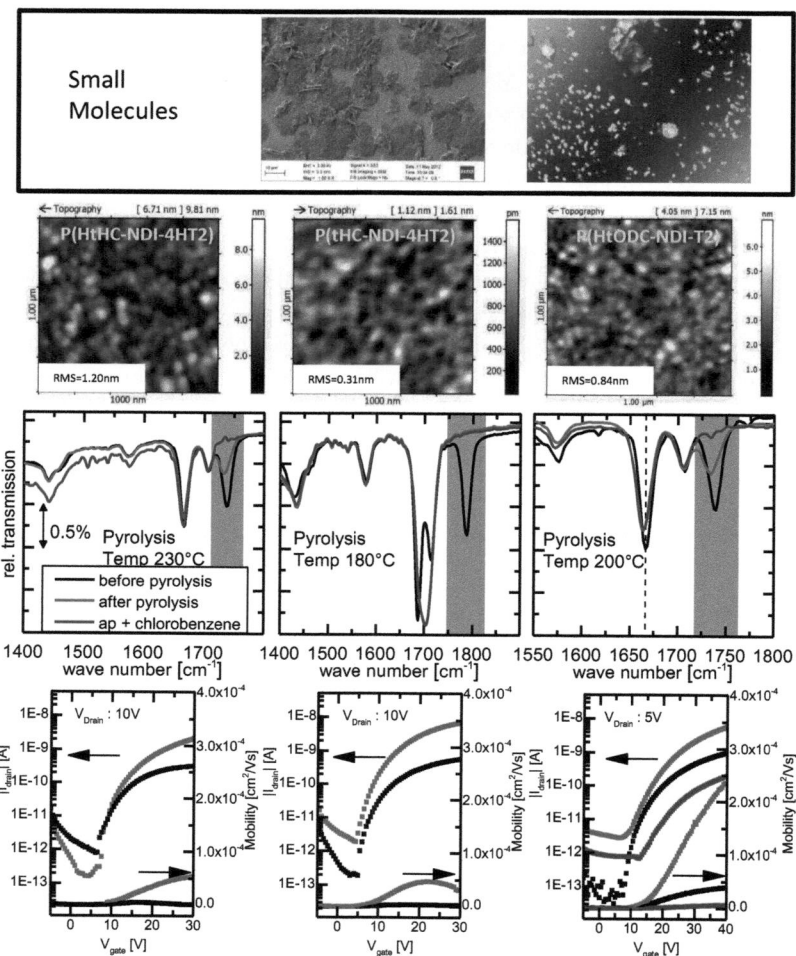

Figure 6.29: Representative data of small molecules (first row) and polymers investigated in this work. First row presents problems with dewetting and inhomogeneity of used small molecules. The second row shows a comparison of AFM data collected from the three investigated polymers. The third and fourth row present the corresponding IR and OFET measurements.

7. Summary

This thesis was carried out within the scope of the MORPHEUS project. Two main topics were investigated in detail. First, the optimization of the organic/inorganic contacts interface was studied in terms of energetic and morphological properties. In the next part possible solution proposals for the intermixing of solution processed organic semiconductor materials were investigated.

In the first part already known materials, which are able to influence the properties of the used electrodes, were studied. In 2012 PEI and PEIE were introduced in the field of organic electronics by Zhou et al. These polymers provide a different approach to alter the properties of the used electrode and are substrate independent in contrast to SAMs. The polymers were also characterized with various analytical methods as well as used as electron injection layers in OSCs in combination with an ITO electrode. The achieved results suggest that the polymers have a great potential to act as interfacial layer to lower the work function of air stable high WF electrodes. Even though the working principle of these polymers is still not fully understood some of the in literature suggested working principles were proven to be false. It turns out that these polymers are most probably not "only" physisorbed on the substrate surface. The XPS and KP measurements performed within this study indicate some chemical reaction between the polymer and the underlying substrate. In collaboration with Stolz et al. the process of electrostatic self-assembly between protonated PEI/PEIE amine groups and oxygen atoms on the electrode surface reported by Kang et al. was proven to be incorrect.

In order to establish an optimized SAM formation process a combination of the following methods was applied in cooperation with research groups at the InnovationLab: KP, PES, IRRAS and device characterization. Commercially available SAMs were characterized and used in organic devices and during preliminary investigations new insights were found. First a facility appropriate

7. Summary

substrate cleaning procedure prior to SAM application was established since the substrates preparation methods proposed by many different sources [33, 40, 82–86] lead to unstable and non-reproducible results. It turned out that simple argon plasma treatment on freshly prepared gold substrates provided surfaces with negligible adsorbate contamination and satisfactory WF values as a SAM treatment starting point. In further investigations on the reference molecule PFDT, it was proven that a deposition of SAM from solution in line with a high throughput printing process is in principle possible. As a result one needs to ensure that crucial process parameters like concentration and immersion time are perfectly matched. Therefore a new parameter was introduced namely exposure, which is described as the product of used molecule concentration and immersion time. A WF shift of $\sim 0.5\,\text{eV}$ was reached within $\sim 10\,\text{s} \cdot \text{mM}$ of exposure. Furthermore, it was shown that processing in ambient atmosphere is also possible, however, the preparation time (mostly immersion time) should be considered as critical in this case. A loss in SAM functionality was presented most probably caused by a change in the molecular dipole induced by oxidative degradation, if longer immersion times ($\geq 5^6\,\text{s}$) were used. Additionally, it was found that contamination of the substrate prepared for SAM treatment is not necessarily hindering the functionality of the monolayer but has a large influence on velocity of the SAM formation process.

In cooperation with the synthesis group of the University of Heidelberg more precisely Malte Jesper new SAM molecules were designed, synthesized and characterized using different analytical methods. Using the knowledge gained from the comprehensive PFDT investigation very similar characterizations were performed on the SAM molecules. The compounds were not only designed to be capable to influence the electronic but also the morphological aspects. Already the preliminary investigations with KP and goniometer setup provided encouraging results. A WF shift of around $\sim 1\,\text{eV}$ and almost doubling the area of the wetting envelope was achieved on substrates treated with Juls SAM. Further investigations using PES and IRRAS techniques confirmed and showed even higher WF shifts for each used molecule but also have unexpectedly shown that an additional degree of freedom in the orientation does not necessarily lead to more upright "standing" molecules. In collaboration with Milan Alt from KIT several batches of OFET devices treated with the new SAM molecules were prepared and characterized. It was shown that the contact resistance derived by the TLM method is almost two orders of magnitude lower (from $8 \cdot 10^{-6}$ to $1.4 \cdot 10^{-5}\,\Omega\text{cm}$) than in the case of untreated OFETs. A very first comparison between the PEIE benchmark system and the most promising SAM provides encouraging results. The same concept was transferred to metal oxides by changing the anchor group of these molecules. First characterizations of the until now only two available SAMs provide satisfactory results, however, the most promising molecule is still unstable.

Summing up the first part, a new class of SAMs was derived to be used in organic electronics which consists of small molecules and therefore allows for unhindered charge transport across the interface and is able to significantly change the work function of noble metals. The application of a treatment with the new molecules leads to a strong wettability improvement of the used substrate and decreases its WF by more than $\sim 1\,\text{eV}$. Moreover a significant increase in device characteristics was achieved, although not as high as in the case of the benchmark polymers. However, taking into account the results of these rather simple molecules, one can imagine various future SAM molecules: molecules with an even higher impact on the WF as well as on the wetting behavior of a given substrate. Further functionalization of the molecules using synthetic approaches would give access to impart crystallization dominating moieties while at the same time controlling the organic-metal interface from the energetic side.

In the second part the new synthesized small molecules and polymers (Torben Adermann, OCI) based on NDI were investigated. The new semiconductor materials ought be processed from solution and after thermal treatment converted to insoluble layers with no or only marginal loss in electrical properties. In the best case the temperature of the pyrolysis process should not exceed 200 °C in order to stay compatible with flexible substrates. Both small molecules and polymers were investigated in detail. Many attempts to prepare homogenous films from the small molecules were carried out, however, the layers were not able to endure the pyrolysis process. Only very rough layers with cracks and defects were obtained and no working devices could be prepared using the small molecules.

On the other hand all investigated polymers fulfilled most of the require-

7. Summary

ments. The IR and XPS measurements performed on P(HtHC-NDI-4HT2) and P(HtODC-NDI-T2) polymers indicate that some residuals of the thermally labile groups are still in the thin film even after the pyrolysis process. However, the achieved solvent resistance is more than sufficient in order to perform further preparation steps. The side chain residuals cause ambiguous effects changing the electrical properties in either way. One possible reason for the negative effect could be the built-up stress in the layer which somehow hinders the transport properties. In both cases a temperature above 200 °C was needed in order to achieve the desired functionality.

All requirements like processability, pyrolysis temperature and solvent resistance after pyrolysis were met only by the P(tHC-NDI-4HT2) polymer. The samples pyrolyzed by 180 °C and investigated by XPS and IR showed no residuals of the solubilizing groups, which means that the pyrolysis process was successful and the cleaved side chains are volatile enough to leave the layer. As expected an excellent solvent resistance and no change in the morphology was measured even after 60 seconds of solvent treatment. The achieved OFET performance, however, was lower than in case of the P(HtODC-NDI-T2) polymer. One has to note that no stack optimizations were carried out in order to improve the device performance. The results presented in this work show that the problems with the process from solution like intermixing of the used materials due to limited amount of orthogonal solvents are resolvable with the proposed approach. In order to provide competitive results further optimizations in chemical structure and on the used OFET stack must to be performed or this approach needs to be transferred to materials with even higher intrinsic mobility.

Bibliography

[1] SHOCKLEY, WILLIAM: *Transistor technology evokes new physics*. Nobel Lecture, pages 344–374, 1956.

[2] POCHETTINO, A.: *Sul comportamento foto-elettrico dell'antracene*. Acad. Lincei Rend, 15, 1906.

[3] POPE, M., H. P. KALLMANN and P. MAGNANTE: *Electroluminescence in Organic Crystals*. The Journal of Chemical Physics, 38(8):2042–2043, April 1963.

[4] SHIRAKAWA, HIDEKI, EDWIN J. LOUIS, ALAN G. MACDIARMID, CHWAN K. CHIANG and ALAN J. HEEGER: *Synthesis of electrically conducting organic polymers: halogen derivatives of polyacetylene, (CH)x*. Journal of the Chemical Society, Chemical Communications, (16):578–580, January 1977.

[5] TANG, C. W. and S. A. VANSLYKE: *Organic electroluminescent diodes*. Applied Physics Letters, 51(12):913–915, September 1987.

[6] HEBNER, T. R., C. C. WU, D. MARCY, M. H. LU and J. C. STURM: *Ink-jet printing of doped polymers for organic light emitting devices*. Applied Physics Letters, 72(5):519–521, February 1998.

[7] *InnovationLab GmbH*. http://www.innovationlab.de.

[8] HELGESEN, MARTIN, ROAR SØNDERGAARD and FREDERIK C. KREBS: *Advanced materials and processes for polymer solar cell devices*. Journal of Materials Chemistry, 20(1):36–60, December 2009.

[9] MORRISON, ROBERT T. and ROBERT N. BOYD: *Organic Chemistry, 6th Edition*. Prentice Hall, Englewood, Cliffs, N.J, 6th edition edition, January 1992.

Bibliography

[10] MCMURRY: *Chemistry Annotated Instructors Edition*.

[11] PAULING, LINUS.: *The Nature of the Chemical Bond. Application of Results Obtained from the Quantum Mechanics and from a Theory of Paramagnetic Susceptibility to the Structure of Molecules*. Journal of the American Chemical Society, 53(4):1367–1400, April 1931.

[12] PETRUCCI, RALPH H., F. GEOFFREY HERRING, JEFFRY D. MADURA and CAREY BISSONNETTE: *General Chemistry: Principles and Modern Applications*. Pearson Prentice Hall, Toronto, Ont., 10th edition edition, May 2010.

[13] WEINHOLD, FRANK: *Valency and Bonding: A Natural Bond Orbital Donor-Acceptor Perspective*. Cambridge University Press, Cambridge, UK ; New York, June 2005.

[14] BÄSSLER, HEINZ and ANNA KÖHLER: *Charge Transport in Organic Semiconductors*. In METZGER, ROBERT M. (editor): *Unimolecular and Supramolecular Electronics I*, number 312 in *Topics in Current Chemistry*, pages 1–65. Springer Berlin Heidelberg, January 2012.

[15] SCHWOERER, MARKUS and HANS CHRISTOPH WOLF: *Organic Molecular Solids*. Wiley-VCH Verlag GmbH & Co. KGaA, Weinheim, Auflage: 1. Auflage edition, December 2006.

[16] FORREST, STEPHEN R.: *The path to ubiquitous and low-cost organic electronic appliances on plastic*. Nature, 428(6986):911–918, April 2004.

[17] KEPLER, R. G.: *Charge Carrier Production and Mobility in Anthracene Crystals*. Physical Review, 119(4):1226–1229, August 1960.

[18] *Physik der Halbleiterbauelemente - Einführendes Lehrbuch für Ingenieure und Physiker*.

[19] NOVIKOV, S. V.: *Hopping charge transport in organic materials*. Russian Journal of Electrochemistry, 48(4):388–400, April 2012.

[20] KAHN, ANTOINE, NORBERT KOCH and WEIYING GAO: *Electronic structure and electrical properties of interfaces between metals and π-conjugated molecular films*. Journal of Polymer Science Part B: Polymer Physics, 41(21):2529–2548, November 2003.

[21] SALANECK, WILLIAM R., KAZUHIKO SEKI and ANTOINE KAHN: *Conjugated Polymer and Molecular Interfaces: Science and Technology for Photonic and Optoelectronic Application: Science and Technology for Photonic and Optoelectronic Applications*. Marcel Dekker Inc, New York, November 2001.

[22] ISHII, HISAO, KIYOSHI SUGIYAMA, EISUKE ITO and KAZUHIKO SEKI: *Energy Level Alignment and Interfacial Electronic Structures at Organic/Metal and Organic/Organic Interfaces*. Advanced Materials, 11(8):605–625, June 1999.

[23] MEYERHOF, WALTER E.: *Contact Potential Difference in Silicon Crystal Rectifiers*. Physical Review, 71(10):727–735, May 1947.

[24] HIMPSEL, F. J., G. HOLLINGER and R. A. POLLAK: *Determination of the Fermi-level pinning position at Si(111) surfaces*. Physical Review B, 28(12):7014–7018, December 1983.

[25] CRISPIN, XAVIER, VICTOR GESKIN, ANNICA CRISPIN, JÉRÔME CORNIL, ROBERTO LAZZARONI, WILLIAM R. SALANECK and JEAN-LUC BRÉDAS: *Characterization of the interface dipole at organic/ metal interfaces*. Journal of the American Chemical Society, 124(27):8131–8141, July 2002.

[26] MUOZ, A., N. CHETTY and RICHARD M. MARTIN: *Modification of heterojunction band offsets by thin layers at interfaces: Role of the interface dipole*. Physical Review B, 41(5):2976–2981, February 1990.

[27] TERSOFF, J.: *Theory of semiconductor heterojunctions: The role of quantum dipoles*. Physical Review B, 30(8):4874–4877, October 1984.

[28] SAIVE, REBECCA: *Investigation of the potential distribution within organic solar cells by scanning Kelvin probe microscopy*. PhD thesis,

Heidelberg, Univ., Diss., 2014, 2014. Zsfassungen in dt. und engl. Sprache.

[29] HALIK, MARCUS and ANDREAS HIRSCH: *The Potential of Molecular Self-Assembled Monolayers in Organic Electronic Devices*. Advanced Materials, 23(22-23):2689–2695, June 2011.

[30] LIAO, KUNG-CHING, AHMAD G. ISMAIL, LAURENT KREPLAK, JEFFREY SCHWARTZ and IAN G. HILL: *Designed Organophosphonate Self-Assembled Monolayers Enhance Device Performance of Pentacene-Based Organic Thin-Film Transistors*. Advanced Materials, 22(28):3081–3085, July 2010.

[31] YAGI, IWAO, KAZUHITO TSUKAGOSHI and YOSHINOBU AOYAGI: *Modification of the electric conduction at the pentacene/SiO2 interface by surface termination of SiO2*. Applied Physics Letters, 86(10):103502, March 2005.

[32] VOS, JOHANNES G., ROBERT J. FORSTER and TIA E. KEYES: *Formation and Characterization of Modified Surfaces*. In *Interfacial Supramolecular Assemblies*, pages 87–152. John Wiley & Sons, Ltd, 2003.

[33] SCHWARTZ, DANIEL K: *Mechanisms and kinetics of self-assembled monolayer formation*. Annual Review of Physical Chemistry, 52(1):107–137, 2001.

[34] AKKERMAN, HYLKE B., RONALD C. G. NABER, BERT JONGBLOED, PAUL A. VAN HAL, PAUL W. M. BLOM, DAGO M. DE LEEUW and BERT DE BOER: *Electron tunneling through alkanedithiol self-assembled monolayers in large-area molecular junctions*. Proceedings of the National Academy of Sciences of the United States of America, 104(27):11161–11166, July 2007.

[35] KRONEMEIJER, A. J., E. H. HUISMAN, H. B. AKKERMAN, A. M. GOOSSENS, I. KATSOURAS, P. A. VAN HAL, T. C. T. GEUNS, S. J. VAN DER MOLEN, P. W. M. BLOM and D. M. DE LEEUW: *Electrical characteristics of conjugated self-assembled monolayers in large-area*

molecular junctions. Applied Physics Letters, 97(17):173302, October 2010.

[36] AKKERMAN, HYLKE B., AUKE J. KRONEMEIJER, PAUL A. VAN HAL, DAGO M. DE LEEUW, PAUL W. M. BLOM and BERT DE BOER: *Self-assembled-monolayer formation of long alkanedithiols in molecular junctions*. Small (Weinheim an Der Bergstrasse, Germany), 4(1):100–104, January 2008.

[37] *An Introduction to Ultrathin Organic Films: From Langmuir–Blodgett to Self–Assembly.*

[38] CALHOUN, M. F., J. SANCHEZ, D. OLAYA, M. E. GERSHENSON and V. PODZOROV: *Electronic functionalization of the surface of organic semiconductors with self-assembled monolayers*. Nature Materials, 7(1):84–89, January 2008.

[39] LAVRICH, DAVID J., SEAN M. WETTERER, STEVEN L. BERNASEK and GIACINTO SCOLES: *Physisorption and Chemisorption of Alkanethiols and Alkyl Sulfides on Au(111)*. The Journal of Physical Chemistry B, 102(18):3456–3465, April 1998.

[40] ULMAN, ABRAHAM: *Formation and structure of self-assembled monolayers*. Chemical reviews, 96(4):1533–1554, 1996.

[41] BENCINI, ALESSANDRO, GOPALAN RAJARAMAN, FEDERICO TOTTI and MATTEO TUSA: *Modeling thiols on Au(111): Structural, thermodynamic and magnetic properties of simple thiols and thiol-radicals*. Superlattices and Microstructures, 46(1–2):4–9, July 2009.

[42] SCHREIBER, FRANK: *Structure and growth of self-assembling monolayers*. Progress in Surface Science, 65(5-8):151–257, November 2000.

[43] DUBOIS, L H and R G NUZZO: *Synthesis, Structure, and Properties of Model Organic Surfaces*. Annual Review of Physical Chemistry, 43(1):437–463, 1992.

[44] JÄNTSCH, O.: *Zur theorie der chemisorption*. Journal of Physics and Chemistry of Solids, 21(1–2):33–39, October 1961.

Bibliography

[45] SCHWABL, FRANZ: *Statistical Mechanics*. Springer, Berlin ; New York, Auflage: 2nd ed. 2006 edition, June 2006.

[46] BAEHR, HANS DIETER: *Thermodynamik: Grundlagen und technische Anwendungen*. Springer, Berlin, Heidelberg, Auflage: 14. Aufl. 2009 edition, August 2009.

[47] KARPOVICH, D. S. and G. J. BLANCHARD: *Direct Measurement of the Adsorption Kinetics of Alkanethiolate Self-Assembled Monolayers on a Microcrystalline Gold Surface*. Langmuir, 10(9):3315–3322, September 1994.

[48] LANGMUIR, IRVING: *Surface Chemistry*. Chemical reviews, 13(2):147–191, 1933.

[49] GODIN, MICHEL, P. J. WILLIAMS, VINCENT TABARD-COSSA, OLIVIER LAROCHE, L. Y. BEAULIEU, R. B. LENNOX and PETER GRÜTTER: *Surface Stress, Kinetics, and Structure of Alkanethiol Self-Assembled Monolayers*. Langmuir, 20(17):7090–7096, August 2004.

[50] BOER, B. DE, A. HADIPOUR, M. M. MANDOC, T. VAN WOUDENBERGH and P. W. M. BLOM: *Tuning of Metal Work Functions with Self-Assembled Monolayers*. Advanced Materials, 17(5):621–625, 2005.

[51] ZEHNER, ROBERT W., BRADLEY F. PARSONS, RICHARD P. HSUNG and LAWRENCE R. SITA: *Tuning the Work Function of Gold with Self-Assembled Monolayers Derived from X-[C6H4-C?C-]nC6H4-SH (n = 0, 1, 2; X = H, F, CH3, CF3, and OCH3)*. Langmuir, 15(4):1121–1127, February 1999.

[52] HONG, JUNG-PYO, AEE-YOUNG PARK, SEONGHOON LEE, JIHOON KANG, NAYOOL SHIN and DO Y YOON: *Tuning of Ag work functions by self-assembled monolayers of aromatic thiols for an efficient hole injection for solution processed triisopropylsilylethynyl pentacene organic thin film transistors*. Applied Physics Letters, 92(14):143311–143311–3, April 2008.

[53] NUZZO, RALPH G., BERNARD R. ZEGARSKI and LAWRENCE H. DUBOIS: *Fundamental studies of the chemisorption of organosulfur compounds on gold(111). Implications for molecular self-assembly on gold surfaces*. Journal of the American Chemical Society, 109(3):733–740, February 1987.

[54] WITTE, GREGOR, SIMON LUKAS, PAUL S. BAGUS and CHRISTOF WÖLL: *Vacuum level alignment at organic/metal junctions: "Cushion" effect and the interface dipole*. Applied Physics Letters, 87(26):263502, December 2005.

[55] ALLOWAY, DANA M., MICHAEL HOFMANN, DARRIN L. SMITH, NADINE E. GRUHN, AMY L. GRAHAM, RAMON COLORADO, VICKI H. WYSOCKI, T. RANDALL LEE, PAUL A. LEE and NEAL R. ARMSTRONG: *Interface Dipoles Arising from Self-Assembled Monolayers on Gold: UV-Photoemission Studies of Alkanethiols and Partially Fluorinated Alkanethiols*. The Journal of Physical Chemistry B, 107(42):11690–11699, 2003.

[56] HEIMEL, GEORG, LORENZ ROMANER, JEAN-LUC BRÉDAS and EGBERT ZOJER: *Interface energetics and level alignment at covalent metal-molecule junctions: pi-conjugated thiols on gold*. Physical Review Letters, 96(19):196806, May 2006.

[57] SUSHKO, MARIA L. and ALEXANDER L. SHLUGER: *Intramolecular Dipole Coupling and Depolarization in Self-Assembled Monolayers*. Advanced Functional Materials, 18(15):2228–2236, August 2008.

[58] RISSNER, FERDINAND, GEROLD M. RANGGER, OLIVER T. HOFMANN, ANNA M. TRACK, GEORG HEIMEL and EGBERT ZOJER: *Understanding the Electronic Structure of Metal/SAM/Organic-Semiconductor Heterojunctions*. ACS Nano, 3(11):3513–3520, November 2009.

[59] KELVIN, LORD: *V. Contact electricity of metals*. Philosophical Magazine Series 5, 46(278):82–120, 1898.

[60] *KP Technology Ltd; http://www.kelvinprobe.info*.

Bibliography

[61] TERADA, MASASHI, NOBUYUKI NAKAMURA, YOICHI NAKAI, YASUYUKI KANAI, SHUNSUKE OHTANI, KEN-ICHIRO KOMAKI and YASUNORI YAMAZAKI: *Observation of an HCI-induced nano-dot on an HOPG surface with STM and AFM.* Nuclear Instruments and Methods in Physics Research Section B: Beam Interactions with Materials and Atoms, 235(1-4):452–455, July 2005.

[62] MANKEL, ERIC: *Elektronische Eigenschaften von Heterosystemen organischer und anorganischer Halbleiter: Präparation, Modifikation und Charakterisierung von Grenzflächen und Kompositen.* Dissertation, TU Darmstadt, July 2011.

[63] KLEIN, ANDREAS, THOMAS MAYER, ANDREAS THISSEN and WOLFRAM JAEGERMANN: *Photoelectron Spectroscopy in Materials Science and Physical Chemistry: Analysis of Composition, Chemical Bonding, and Electronic Structure of Surfaces and Interfaces.* In SCHÄFER, ROLF and PETER C. SCHMIDT (editors): *Methods in Physical Chemistry*, pages 477–512. Wiley-VCH Verlag GmbH & Co. KGaA, 2012.

[64] MANKEL, ERIC: *Grundlagen der Röntgenphotoelektronenspektroskopie, Praktikumsversuch Bachelor 5. Semester, WS 2011/2012.* FB Material-Geowissenschaften, FG Oberflächenforschung, TU-Darmstadt, 2011.

[65] EINSTEIN, A.: *Über einen die Erzeugung und Verwandlung des Lichtes betreffenden heuristischen Gesichtspunkt.* Annalen der Physik, 322(6):132–148, January 1905.

[66] GAO, YONGLI: *Surface analytical studies of interfaces in organic semiconductor devices.* Materials Science and Engineering: R: Reports, 68(3):39–87, April 2010.

[67] SEAH, M. P. and W. A. DENCH: *Quantitative electron spectroscopy of surfaces: A standard data base for electron inelastic mean free paths in solids.* Surface and Interface Analysis, 1(1):2–11, February 1979.

[68] MAIBACH, JULIA: *Preparation and Characterization of Solution Processed Organic Semiconductor Interfaces: Electronic Properties of*

Thiophene-Fullerene based Donor-Acceptor Systems. Dissertation, TU Darmstadt, Darmstadt, 2014.

[69] STENZEL, OLAF: *Das Dünnschichtspektrum*. Wiley-VCH Verlag GmbH, Berlin, March 1996.

[70] HUNKLINGER, SIEGFRIED: *Festkörperphysik*. Oldenbourg Wissenschaftsverlag, München; Wien, October 2007.

[71] BRENDEL, R. and D. BORMANN: *An infrared dielectric function model for amorphous solids*. Journal of Applied Physics, 71(1):1–6, January 1992.

[72] LEHMANN, A.: *Theory of Infrared Transmission Spectra of Thin Insulating Films*. physica status solidi (b), 148(1):401–405, July 1988.

[73] BERREMAN, D. W.: *Infrared Absorption at Longitudinal Optic Frequency in Cubic Crystal Films*. Physical Review, 130(6):2193–2198, June 1963.

[74] GLASER, TOBIAS: *Infrarotspektroskopische Untersuchung der p-Dotierung organischer Halbleiter mit Übergangsmetalloxiden*. Dissertation, 2013.

[75] HILLEBRANDT, SABINA: *Infrarot-Reflexions-Absorptions-Spektroskopie an selbstorganisierenden Monolagen auf Gold*. Masterarbeit, Ruprecht-Karls-Universität, Heidelberg, 2014.

[76] MCINTYRE, J. D. E. and D. E. ASPNES: *Differential reflection spectroscopy of very thin surface films*. Surface Science, 24(2):417–434, February 1971.

[77] GREENLER, ROBERT G.: *Infrared Study of Adsorbed Molecules on Metal Surfaces by Reflection Techniques*. The Journal of Chemical Physics, 44(1):310–315, January 1966.

[78] SCHRADER, BERNHARD: *Infrared and Raman Spectroscopy: Methods and Applications*. Wiley-VCH, Weinheim ; New York, April 1995.

Bibliography

[79] BECK, SEBASTIAN: *Infrarotspektroskopie an organischen Charge-Transfer-Komplexen*. Diplomarbeit, Ruprecht-Karls-Universität, Heidelberg, 2011.

[80] BINNIG, G., C. F. QUATE and CH. GERBER: *Atomic Force Microscope*. Physical Review Letters, 56(9):930–933, March 1986.

[81] *Hersteller von Rastersondenmikroskopen, Rasterkraftmikroskopen, Rastertunnelmikroskopen;* http://www.dme-spm.de.

[82] WHITESIDES, GEORGE M, JENNAH K KRIEBEL and J CHRISTOPHER LOVE: *Molecular engineering of surfaces using self-assembled monolayers*. Science progress, 88(Pt 1):17–48, 2005.

[83] BAIN, COLIN D., E. BARRY TROUGHTON, YU TAI TAO, JOSEPH EVALL, GEORGE M. WHITESIDES and RALPH G. NUZZO: *Formation of monolayer films by the spontaneous assembly of organic thiols from solution onto gold*. Journal of the American Chemical Society, 111(1):321–335, January 1989.

[84] STETTNER, JOHANNA: *Self assembled monolayer formation of alkanethiols on gold: Growth from solution versus*.

[85] VERICAT, C., M. E. VELA and R. C. SALVAREZZA: *Self-assembled monolayers of alkanethiols on Au(111): surface structures, defects and dynamics*. Physical Chemistry Chemical Physics, 7(18):3258–3268, August 2005.

[86] BISHOP, ADEANA R and RALPH G NUZZO: *Self-assembled monolayers: Recent developments and applications*. Current Opinion in Colloid & Interface Science, 1(1):127–136, February 1996.

[87] KAWANO, HIROYUKI: *Effective work functions for ionic and electronic emissions from mono- and polycrystalline surfaces*. Progress in Surface Science, 83(1–2):1–165, February 2008.

[88] RON, HANNOCH and ISRAEL RUBINSTEIN: *Alkanethiol Monolayers on Preoxidized Gold. Encapsulation of Gold Oxide under an Organic Monolayer*. Langmuir, 10(12):4566–4573, December 1994.

Bibliography

[89] BAGUS, PAUL S., VOLKER STAEMMLER and CHRISTOF WÖLL: *Exchangelike Effects for Closed-Shell Adsorbates: Interface Dipole and Work Function.* Physical Review Letters, 89(9):096104, August 2002.

[90] HWANG, JAEHYUNG, ALAN WAN and ANTOINE KAHN: *Energetics of metal–organic interfaces: New experiments and assessment of the field.* Materials Science and Engineering: R: Reports, 64(1–2):1–31, March 2009.

[91] HÜCKSTÄDT, C., S. SCHMIDT, S. HÜFNER, F. FORSTER, F. REINERT and M. SPRINGBORG: *Work function studies of rare-gas/noble metal adsorption systems using a Kelvin probe.* Physical Review B, 73(7):075409, February 2006.

[92] DE RENZI, V., R. ROUSSEAU, D. MARCHETTO, R. BIAGI, S. SCANDOLO and U. DEL PENNINO: *Metal work-function changes induced by organic adsorbates: a combined experimental and theoretical study.* Physical Review Letters, 95(4):046804, July 2005.

[93] HÄNSEL, MARC: *Präparation und Charakterisierung selbstorganisierender Monolagen auf polykristallinen Substraten.* Masterarbeit, Ruprecht-Karls-Universität, Heidelberg, 2013.

[94] MURDEY, RICHARD J. and WILLIAM R. SALANECK: *Charge Injection Barrier Heights Across Multilayer Organic Thin Films.* Japanese Journal of Applied Physics, 44(6R):3751, June 2005.

[95] CLARK, MICHAEL D. and BENJAMIN J. LEEVER: *Analysis of ITO cleaning protocol on surface properties and polymer: Fullerene bulk heterojunction solar cell performance.* Solar Energy Materials and Solar Cells, 116:270–274, September 2013.

[96] CHOCKALINGAM, MUTHUKUMAR, NADIM DARWISH, GUILLAUME LE SAUX and J. JUSTIN GOODING: *Importance of the Indium Tin Oxide Substrate on the Quality of Self-Assembled Monolayers Formed from Organophosphonic Acids.* Langmuir, 27(6):2545–2552, March 2011.

Bibliography

[97] SONG, MYUNGKWAN, JAE-WOOK KANG, DONG-HO KIM, JUNG-DAE KWON, SUNG-GYU PARK, SANGGIL NAM, SUNGJIN JO, SEUNG YOON RYU and CHANG SU KIM: *Self-assembled monolayer as an interfacial modification material for highly efficient and air-stable inverted organic solar cells.* Applied Physics Letters, 102(14):143303–143303-5, April 2013.

[98] HOTCHKISS, PETER J., SIMON C. JONES, SERGIO A. PANIAGUA, ASHA SHARMA, BERNARD KIPPELEN, NEAL R. ARMSTRONG and SETH R. MARDER: *The Modification of Indium Tin Oxide with Phosphonic Acids: Mechanism of Binding, Tuning of Surface Properties, and Potential for Use in Organic Electronic Applications.* Accounts of Chemical Research, 45(3):337–346, March 2012.

[99] ZHOU, YINHUA, CANEK FUENTES-HERNANDEZ, JAEWON SHIM, JENS MEYER, ANTHONY J. GIORDANO, HONG LI, PAUL WINGET, THEODOROS PAPADOPOULOS, HYEUNSEOK CHEUN, JUNGBAE KIM, MATHIEU FENOLL, AMIR DINDAR, WOJCIECH HASKE, EHSAN NAJAFABADI, TALHA M. KHAN, HOSSEIN SOJOUDI, STEPHEN BARLOW, SAMUEL GRAHAM, JEAN-LUC BRÉDAS, SETH R. MARDER, ANTOINE KAHN and BERNARD KIPPELEN: *A Universal Method to Produce Low–Work Function Electrodes for Organic Electronics.* Science, 336(6079):327–332, April 2012.

[100] WURFEL, PETER: *Physics of Solar Cells: From Basic Principles to Advanced Concepts.* Wiley-VCH, Weinheim, Auflage: 2. aktualis. u. erg. Auflage edition, April 2009.

[101] HUNKLINGER, SIEGFRIED: *Festkörperphysik.* Oldenbourg Wissenschaftsverlag, Muünchen; Wien, October 2007.

[102] FRENKEL, J.: *On the Transformation of light into Heat in Solids. I.* Physical Review, 37(1):17–44, January 1931.

[103] ASHCROFT, NEIL W. and N. DAVID MERMIN: *Solid State Physics.* Cengage Learning Emea, New York, January 1976.

Bibliography

[104] BRUTTING, WOLFGANG and WALTER RIESS: *Grundlagen der organischen Halbleiter*. Physik Journal, 7(5):33, 2008.

[105] PARK, SUNG HEUM, ANSHUMAN ROY, SERGE BEAUPRÉ, SHINUK CHO, NELSON COATES, JI SUN MOON, DANIEL MOSES, MARIO LECLERC, KWANGHEE LEE and ALAN J. HEEGER: *Bulk heterojunction solar cells with internal quantum efficiency approaching 100%*. Nature Photonics, 3(5):297–302, May 2009.

[106] MÜLLER, CHRISTIAN: *Raster-Kelvin-Mikroskopie an Querschnitten organischer Solarzellen*. Masterarbeit, Ruprecht-Karls-Universität, Heidelberg, 2013.

[107] DOU, LETIAN, JINGBI YOU, ZIRUO HONG, ZHENG XU, GANG LI, ROBERT A. STREET and YANG YANG: *25th Anniversary Article: A Decade of Organic/Polymeric Photovoltaic Research*. Advanced Materials, 25(46):6642–6671, December 2013.

[108] DAUME, DOMINIK: *C-V-Messungen an P3HT- und PCBM-Einschichtbauteilen sowie P3HT:PCBM-Doppelschicht- und Bulk-Heterojunction-Solarzellen*. Masterarbeit, Ruprecht-Karls-Universität, Heidelberg, 2012.

[109] NEWMAN, CHRISTOPHER R., C. DANIEL FRISBIE, DEMETRIO A. DA SILVA FILHO, JEAN-LUC BRÉDAS, PAUL C. EWBANK and KENT R. MANN: *Introduction to Organic Thin Film Transistors and Design of n-Channel Organic Semiconductors*. Chem. Mater., 16(23):4436–4451, 2004.

[110] SIRRINGHAUS, HENNING: *25th Anniversary Article: Organic Field-Effect Transistors: The Path Beyond Amorphous Silicon*. Advanced Materials, page n/a–n/a, 2014.

[111] HOROWITZ, GILLES: *Organic Field-Effect Transistors*. Advanced Materials, 10(5):365–377, March 1998.

[112] CELLE, C., C. SUSPÈNE, M. TERNISIEN, S. LENFANT, D. GUÉRIN, K. SMAALI, K. LMIMOUNI, J. P. SIMONATO and D. VUILLAUME: *Interface dipole: Effects on threshold voltage and mobility for both amorphous*

Bibliography

and poly-crystalline organic field effect transistors. Organic Electronics, 15(3):729–737, March 2014.

[113] NATALI, DARIO and MARIO CAIRONI: *Charge Injection in Solution-Processed Organic Field-Effect Transistors: Physics, Models and Characterization Methods.* Advanced Materials, 24(11):1357–1387, March 2012.

[114] TOKITO, SHIZUO, KOJI NODA and YASUNORI TAGA: *Metal oxides as a hole-injecting layer for an organic electroluminescent device.* Journal of Physics D: Applied Physics, 29(11):2750, November 1996.

[115] CHEUN, HYEUNSEOK, CANEK FUENTES-HERNANDEZ, YINHUA ZHOU, WILLIAM J. POTSCAVAGE, SUNG-JIN KIM, JAEWON SHIM, AMIR DINDAR and BERNARD KIPPELEN: *Electrical and Optical Properties of ZnO Processed by Atomic Layer Deposition in Inverted Polymer Solar Cells†.* The Journal of Physical Chemistry C, 114(48):20713–20718, December 2010.

[116] CAMPBELL, I. H., S. RUBIN, T. A. ZAWODZINSKI, J. D. KRESS, R. L. MARTIN, D. L. SMITH, N. N. BARASHKOV and J. P. FERRARIS: *Controlling Schottky energy barriers in organic electronic devices using self-assembled monolayers.* Physical Review B, 54(20):R14321–R14324, November 1996.

[117] XIONG, TAO, FENGXIA WANG, XIANFENG QIAO and DONGGE MA: *A soluble nonionic surfactant as electron injection material for high-efficiency inverted bottom-emission organic light emitting diodes.* APL: Organic Electronics and Photonics, 1(9):352–352, September 2008.

[118] SARACCO, EMELINE, BENJAMIN BOUTHINON, JEAN-MARIE VERILHAC, CAROLINE CELLE, NICOLAS CHEVALIER, DENIS MARIOLLE, OLIVIER DHEZ and JEAN-PIERRE SIMONATO: *Work Function Tuning for High-Performance Solution-Processed Organic Photodetectors with Inverted Structure.* Advanced Materials, 25(45):6534–6538, December 2013.

[119] KYAW, AUNG KO KO, DONG HWAN WANG, VINAY GUPTA, JIE ZHANG, SURESH CHAND, GUILLERMO C. BAZAN and ALAN J. HEEGER: *Efficient Solution-Processed Small-Molecule Solar Cells with Inverted Structure.* Advanced Materials, 25(17):2397–2402, May 2013.

[120] KIM, YOUNG-HOON, TAE-HEE HAN, HIMCHAN CHO, SUNG-YONG MIN, CHANG-LYOUL LEE and TAE-WOO LEE: *Polyethylene Imine as an Ideal Interlayer for Highly Efficient Inverted Polymer Light-Emitting Diodes.* Advanced Functional Materials, 24(24):3808–3814, June 2014.

[121] KANG, HONGKYU, SOONIL HONG, JONGJIN LEE and KWANGHEE LEE: *Electrostatically Self-Assembled Nonconjugated Polyelectrolytes as an Ideal Interfacial Layer for Inverted Polymer Solar Cells.* Advanced Materials, 24(22):3005–3009, 2012.

[122] FABIANO, SIMONE, SLAWOMIR BRAUN, XIANJIE LIU, ERIC WEVERBERGHS, PASCAL GERBAUX, MATS FAHLMAN, MAGNUS BERGGREN and XAVIER CRISPIN: *Poly(ethylene imine) Impurities Induce n-doping Reaction in Organic (semi)Conductors.* Advanced Materials, July 2014.

[123] STOLZ, SEBASTIAN, MICHAEL SCHERER, ERIC MANKEL, ROBERT LOVRINČIĆ, JANUSZ SCHINKE, WOLFGANG KOWALSKY, WOLFRAM JAEGERMANN, ULI LEMMER, NORMAN MECHAU and GERARDO HERNANDEZ-SOSA: *Investigation of Solution-Processed Ultrathin Electron Injection Layers for Organic Light-Emitting Diodes.* ACS Applied Materials & Interfaces, 6(9):6616–6622, May 2014.

[124] GÜNES, SERAP, HELMUT NEUGEBAUER and NIYAZI SERDAR SARICIFTCI: *Conjugated polymer-based organic solar cells.* Chemical Reviews, 107(4):1324–1338, April 2007.

[125] KRÖGER, M., S. HAMWI, J. MEYER, T. RIEDL, W. KOWALSKY and A. KAHN: *Role of the deep-lying electronic states of MoO3 in the enhancement of hole-injection in organic thin films.* Applied Physics Letters, 95(12):123301–123301–3, September 2009.

[126] GUAN, ZE-LEI, JONG BOK KIM, HE WANG, CHERNO JAYE, DANIEL A. FISCHER, YUEH-LIN LOO and ANTOINE KAHN:

Direct determination of the electronic structure of the poly(3-hexylthiophene):phenyl-[6,6]-C61 butyric acid methyl ester blend. Organic Electronics, 11(11):1779–1785, November 2010.

[127] SCHLAF, R., B. A. PARKINSON, P. A. LEE, K. W. NEBESNY, G. JABBOUR, B. KIPPELEN, N. PEYGHAMBARIAN and N. R. ARMSTRONG: *Photoemission spectroscopy of LiF coated Al and Pt electrodes.* Journal of Applied Physics, 84(12):6729–6736, December 1998.

[128] SHAHEEN, S. E., G. E. JABBOUR, M. M. MORRELL, Y. KAWABE, B. KIPPELEN, N. PEYGHAMBARIAN, M.-F. NABOR, R. SCHLAF, E. A. MASH and N. R. ARMSTRONG: *Bright blue organic light-emitting diode with improved color purity using a LiF/Al cathode.* Journal of Applied Physics, 84(4):2324–2327, August 1998.

[129] LOVE, J. CHRISTOPHER, LARA A. ESTROFF, JENNAH K. KRIEBEL, RALPH G. NUZZO and GEORGE M. WHITESIDES: *Self-Assembled Monolayers of Thiolates on Metals as a Form of Nanotechnology.* Chemical Reviews, 105(4):1103–1170, April 2005.

[130] WHITESIDES, GEORGE M and BARTOSZ GRZYBOWSKI: *Self-Assembly at All Scales.* Science, 295(5564):2418–2421, March 2002.

[131] BIGELOW, W. C., D. L. PICKETT and W. A. ZISMAN: *Oleophobic monolayers: I. Films adsorbed from solution in non-polar liquids.* Journal of Colloid Science, 1(6):513–538, December 1946.

[132] HEIMEL, GEORG, LORENZ ROMANER, EGBERT ZOJER and JEAN-LUC BREDAS: *The Interface Energetics of Self-Assembled Monolayers on Metals.* Accounts of Chemical Research, 41(6):721–729, June 2008.

[133] EVANS, STEPHEN D. and ABRAHAM ULMAN: *Surface potential studies of alkyl-thiol monolayers adsorbed on gold.* Chemical Physics Letters, 170(5–6):462–466, July 1990.

[134] VENKATARAMAN, NAGAIYANALLUR V., STEFAN ZÜRCHER, ANTONELLA ROSSI, SEUNGHWAN LEE, NICOLA NAUJOKS and

NICHOLAS D. SPENCER: *Spatial Tuning of the Metal Work Function by Means of Alkanethiol and Fluorinated Alkanethiol Gradients.* The Journal of Physical Chemistry C, 113(14):5620–5628, April 2009.

[135] IWAMI, YASUNOBU, DAISUKE HOBARA, MASAHIRO YAMAMOTO and TAKASHI KAKIUCHI: *Determination of the potential of zero charge of Au(1 1 1) electrodes modified with thiol self-assembled monolayers using a potential-controlled sessile drop method.* Journal of Electroanalytical Chemistry, 564:77–83, March 2004.

[136] ASADI, KAMAL, FATEMEH GHOLAMREZAIE, EDSGER C. P. SMITS, PAUL W. M. BLOM and BERT DE BOER: *Manipulation of charge carrier injection into organic field-effect transistors by self-assembled monolayers of alkanethiols.* Journal of Materials Chemistry, 17(19):1947–1953, May 2007.

[137] THOMAS, ROSS C., LI. SUN, RICHARD M. CROOKS and ANTONIO J. RICCO: *Real-time measurements of the gas-phase adsorption of n-alkylthiol mono- and multilayers on gold.* Langmuir, 7(4):620–622, April 1991.

[138] KARPOVICH, D. S. and G. J. BLANCHARD: *Direct Measurement of the Adsorption Kinetics of Alkanethiolate Self-Assembled Monolayers on a Microcrystalline Gold Surface.* Langmuir, 10(9):3315–3322, September 1994.

[139] KIM, DONG HO, JAE GWON NO, MASAHIKO HARA and HYE WON LEE: *An Adsorption Process Study on the Self-Assembled Monolayer Formation of Octadecanethiol Chemisorged on Gold Surface.* Bulletin of the Korean Chemical Society, 22(3):276–280, 2001.

[140] BENSEBAA, FARID, RALUCA VOICU, LAURENT HURON, THOMAS H. ELLIS and ERIK KRUUS: *Kinetics of Formation of Long-Chain n-Alkanethiolate Monolayers on Polycrystalline Gold.* Langmuir, 13(20):5335–5340, October 1997.

[141] POIRIER, G. E.: *Coverage-Dependent Phases and Phase Stability of Decanethiol on Au(111).* Langmuir, 15(4):1167–1175, February 1999.

Bibliography

[142] HUTCHINS, DANIEL O., ORB ACTON, TOBIAS WEIDNER, NATHAN CERNETIC, JOE E. BAIO, DAVID G. CASTNER, HONG MA and ALEX K. Y. JEN: *Solid-state densification of spun-cast self-assembled monolayers for use in ultra-thin hybrid dielectrics.* Applied Surface Science, 261:908–915, November 2012.

[143] ACTON, ORB, MANISH DUBEY, TOBIAS WEIDNER, KEVIN M. O'MALLEY, TAE-WOOK KIM, GUY G. TING, DANIEL HUTCHINS, J. E. BAIO, TRACY C. LOVEJOY, ALEXANDER H. GAGE, DAVID G. CASTNER, HONG MA and ALEX K.-Y. JEN: *Simultaneous Modification of Bottom-Contact Electrode and Dielectric Surfaces for Organic Thin-Film Transistors Through Single-Component Spin-Cast Monolayers.* Advanced Functional Materials, 21(8):1476–1488, April 2011.

[144] NIE, HENG-YONG, MARY J. WALZAK and N. STEWART MCINTYRE: *Delivering octadecylphosphonic acid self-assembled monolayers on a Si wafer and other oxide surfaces.* The Journal of Physical Chemistry. B, 110(42):21101–21108, October 2006.

[145] BAIN, COLIN D. and GEORGE M. WHITESIDES: *A study by contact angle of the acid-base behavior of monolayers containing .omega.-mercaptocarboxylic acids adsorbed on gold: an example of reactive spreading.* Langmuir, 5(6):1370–1378, November 1989.

[146] INA RIANASARI, LORENZ WALDER: *Inkjet-printed thiol self-assembled monolayer structures on gold: quality control and microarray electrode fabrication.* Langmuir : the ACS journal of surfaces and colloids, 24(16):9110–7, 2008.

[147] HAMADANI, B. H., D. A. CORLEY, J. W. CISZEK, J. M. TOUR and D. NATELSON: *Controlling Charge Injection in Organic Field-Effect Transistors Using Self-Assembled Monolayers.* Nano Letters, 6(6):1303–1306, June 2006.

[148] BOCK, C., D. V. PHAM, U. KUNZE, D. KÄFER, G. WITTE and CH WÖLL: *Improved morphology and charge carrier injection in pentacene field-effect transistors with thiol-treated electrodes.* Journal of Applied Physics, 100(11):114517, December 2006.

[149] BOUDINET, DAMIEN, MOHAMED BENWADIH, YABING QI, STÉPHANE ALTAZIN, JEAN-MARIE VERILHAC, MICHAEL KROGER, CHRISTOPHE SERBUTOVIEZ, ROMAIN GWOZIECKI, ROMAIN COPPARD, GILLES LE BLEVENNEC, ANTOINE KAHN and GILLES HOROWITZ: *Modification of gold source and drain electrodes by self-assembled monolayer in staggered n- and p-channel organic thin film transistors*. Organic Electronics, 11(2):227–237, February 2010.

[150] HIETZSCHOLD, SEBASTIAN: *Optimierung von Grenzflächen in der organischen Elektronik durch selbstorganisierende Monolagen*. Bachelorarbeit, Ruprecht-Karls-Universität, Heidelberg, 2011.

[151] LAIBINIS, PAUL E., GEORGE M. WHITESIDES, DAVID L. ALLARA, YU TAI TAO, ATUL N. PARIKH and RALPH G. NUZZO: *Comparison of the structures and wetting properties of self-assembled monolayers of n-alkanethiols on the coinage metal surfaces, copper, silver, and gold*. Journal of the American Chemical Society, 113(19):7152–7167, September 1991.

[152] MATTIUZZI, ALICE, IVAN JABIN, CLAIRE MANGENEY, CLÉMENT ROUX, OLIVIA REINAUD, LUIS SANTOS, JEAN-FRANÇOIS BERGAMINI, PHILIPPE HAPIOT and CORINNE LAGROST: *Electrografting of calix[4]arenediazonium salts to form versatile robust platforms for spatially controlled surface functionalization*. Nature Communications, 3:1130, October 2012.

[153] TSENG, CHUNG-TING, YU-HUNG CHENG, MING-CHANG M. LEE, CHIEN-CHUNG HAN, CHIEN-HONG CHENG and YU-TAI TAO: *Study of anode work function modified by self-assembled monolayers on pentacene/fullerene organic solar cells*. Applied Physics Letters, 91(23):233510, December 2007.

[154] LANDEMARK, ERIK, C. J. KARLSSON, Y.-C. CHAO and R. I. G. UHRBERG: *Core-level spectroscopy of the clean Si(001) surface: Charge transfer within asymmetric dimers of the 2×1 and c(4×2) reconstructions*. Physical Review Letters, 69(10):1588–1591, September 1992.

Bibliography

[155] MOULDER, JOHN F., WILLIAM F. STICKLE, PETER E. SOBOL and KENNETH D. BOMBEN: *Handbook of X Ray Photoelectron Spectroscopy: A Reference Book of Standard Spectra for Identification and Interpretation of Xps Data*. Physical Electronics, Eden Prairie, Minn., Auflage: Reissue edition, February 1995.

[156] ALT, M., J. SCHINKE, S. HILLEBRANDT, M. HÄNSEL, G. HERNANDEZ-SOSA, N. MECHAU, T. GLASER, E. MANKEL, M. HAMBURGER, K. DEING, W. JAEGERMANN, A. PUCCI, W. KOWALSKY, U. LEMMER and R. LOVRINCIC: *Processing follows function: Pushing the formation of self-assembled monolayers to high throughput compatible timescales*. Submitted to ACS Nano Interfaces.

[157] ALVES, CARLA A. and MARC D. PORTER: *Atomic force microscopic characterization of a fluorinated alkanethiolate monolayer at gold and correlations to electrochemical and infrared reflection spectroscopic structural descriptions*. Langmuir, 9(12):3507–3512, December 1993.

[158] CHIDSEY, CHRISTOPHER ED and DOMINIC N. LOIACONO: *Chemical functionality in self-assembled monolayers: structural and electrochemical properties*. Langmuir, 6(3):682–691, 1990.

[159] LIU, GANG-YU, PAUL FENTER, CHRISTOPHER E. D. CHIDSEY, D. FRANK OGLETREE, PETER EISENBERGER and MIQUEL SALMERON: *An unexpected packing of fluorinated n-alkane thiols on Au(111): A combined atomic force microscopy and x-ray diffraction study*. The Journal of Chemical Physics, 101(5):4301–4306, September 1994.

[160] TSAI, WEN-TIEN: *Environmental risk assessment of hydrofluoroethers (HFEs)*. Journal of Hazardous Materials, 119(1-3):69–78, March 2005.

[161] PARSONS, JOHN R., MONICA SÁEZ, JAN DOLFING and PIM DE VOOGT: *Biodegradation of perfluorinated compounds*. Reviews of Environmental Contamination and Toxicology, 196:53–71, 2008.

[162] ZHANG, WEIMIN, JEREMY SMITH, RICK HAMILTON, MARTIN HEENEY, JAMES KIRKPATRICK, KIGOOK SONG, SCOTT E. WATKINS, THOMAS ANTHOPOULOS and IAIN MCCULLOCH: *Systematic Improvement in*

Charge Carrier Mobility of Air Stable Triarylamine Copolymers. Journal of the American Chemical Society, 131(31):10814–10815, August 2009.

[163] JESPER, MALTE: *Modifikation von Elektrodenoberflächen durch selbst assoziierte Monolagen :Dipolare Thiole auf Gold.* Masterarbeit, Ruprecht-Karls-Universität, Heidelberg, 2012.

[164] BUTENSCHÖN, HOLGER, K. PETER C. VOLLHARDT and NEIL E. SCHORE: *Organische Chemie.* Wiley-VCH Verlag GmbH & Co. KGaA, Weinheim, Auflage: 5. Auflage edition, November 2011.

[165] W. A. ZISMAN: *Relation of the Equilibrium Contact Angle to Liquid and Solid Constitution.* In *Contact Angle, Wettability, and Adhesion*, volume 43 of *Advances in Chemistry*, pages 1–51. AMERICAN CHEMICAL SOCIETY, January 1964.

[166] BONN, DANIEL, JENS EGGERS, JOSEPH INDEKEU, JACQUES MEUNIER and ETIENNE ROLLEY: *Wetting and spreading.* Reviews of Modern Physics, 81(2):739–805, May 2009.

[167] YOUNG, THOMAS: *An Essay on the Cohesion of Fluids.* Philosophical Transactions of the Royal Society of London, 95:65–87, January 1805.

[168] WOLFRAM, E.: *Adhäsion von Flüssigkeiten an Kunststoffoberflächen.* Kolloid-Zeitschrift und Zeitschrift für Polymere, 182(1-2):75–85, May 1962.

[169] KAELBLE, D. H.: *Dispersion-Polar Surface Tension Properties of Organic Solids.* The Journal of Adhesion, 2(2):66–81, April 1970.

[170] OWENS, D. K. and R. C. WENDT: *Estimation of the surface free energy of polymers.* Journal of Applied Polymer Science, 13(8):1741–1747, August 1969.

[171] RABEL, W: *Einige Aspekte der Benetzungstheorie und ihre Anwendung auf die Untersuchung und Veränderung der Oberflächeneigenschaften von Polymeren.* Farbe und Lacke, 77(10):997–1005, 1971.

Bibliography

[172] MATHIES, FLORIAN: *Ink-Jet Formulierungen auf Basis nichthalogenierter Lösemittel für gedruckte organische Solarzellen*. Bachelorarbeit, Ruprecht-Karls-Universität, Heidelberg, 2011.

[173] KRESS, JOSHUA: *Infrarotspektroskopie an selbstorganisierenden Monolagen auf Gold*. Bachelorarbeit, Ruprecht-Karls-Universität, Heidelberg, 2014.

[174] YAN, HE, ZHIHUA CHEN, YAN ZHENG, CHRISTOPHER NEWMAN, JORDAN R. QUINN, FLORIAN DÖTZ, MARCEL KASTLER and ANTONIO FACCHETTI: *A high-mobility electron-transporting polymer for printed transistors*. Nature, 457(7230):679–686, February 2009.

[175] KLAUK, HAGEN, GÜNTER SCHMID, WOLFGANG RADLIK, WERNER WEBER, LISONG ZHOU, CHRIS D SHERAW, JONATHAN A NICHOLS and THOMAS N JACKSON: *Contact resistance in organic thin film transistors*. Solid-State Electronics, 47(2):297–301, February 2003.

[176] KONDO, TOSHIHIRO, MIWA TAKECHI, YUKARI SATO and KOHEI UOSAKI: *Adsorption behavior of functionalized ferrocenylalkane thiols and disulfide onto Au and ITO and electrochemical properties of modified electrodes: Effects of acyl and alkyl groups attached to the ferrocene ring*. Journal of Electroanalytical Chemistry, 381(1–2):203–209, January 1995.

[177] GARDNER, TIMOTHY J., C. DANIEL FRISBIE and MARK S. WRIGHTON: *Systems for orthogonal self-assembly of electroactive monolayers on Au and ITO: an approach to molecular electronics*. Journal of the American Chemical Society, 117(26):6927–6933, July 1995.

[178] YAN, C., M. ZHARNIKOV, A. GÖLZHÄUSER and M. GRUNZE: *Preparation and Characterization of Self-Assembled Monolayers on Indium Tin Oxide*. Langmuir, 16(15):6208–6215, July 2000.

[179] GAO, WEI, LUCY DICKINSON, CHRISTINA GROZINGER, FREDERICK G. MORIN and LINDA REVEN: *Self-Assembled Monolayers of Alkylphosphonic Acids on Metal Oxides*. Langmuir, 12(26):6429–6435, January 1996.

[180] BROVELLI, DOROTHEE, GEORG HÄHNER, LAURENCE RUIZ, ROLF HOFER, GEROLF KRAUS, ADRIAN WALDNER, JOHANNA SCHLÖSSER, PETER OROSZLAN, MARKUS EHRAT and NICHOLAS D. SPENCER: *Highly Oriented, Self-Assembled Alkanephosphate Monolayers on Tantalum(V) Oxide Surfaces.* Langmuir, 15(13):4324–4327, June 1999.

[181] PANIAGUA, SERGIO A., PETER J. HOTCHKISS, SIMON C. JONES, SETH R. MARDER, ANOMA MUDALIGE, F. SANEEHA MARRIKAR, JEANNE E. PEMBERTON and NEAL R. ARMSTRONG: *Phosphonic Acid Modification of Indium-Tin Oxide Electrodes: Combined XPS/UPS/Contact Angle Studies†.* The Journal of Physical Chemistry C, 112(21):7809–7817, May 2008.

[182] BREWER, SCOTT H., DEREK A. BROWN and STEFAN FRANZEN: *Formation of Thiolate and Phosphonate Adlayers on Indium-Tin Oxide: Optical and Electronic Characterization.* Langmuir, 18(18):6857–6865, September 2002.

[183] BWALYA, RAMOS-DAVID: *Engineering Interfacial Properties Of Inverted Polymer Solar Cells Trough Surface Functionalization.* Bachelorarbeit, Ruprecht-Karls-Universität, Heidelberg, 2014.

[184] ADERMANN, TORBEN: *Organische n-Typ-Halbleitermaterialien mit thermisch labilen löslichkeitsvermittelnden Gruppen.* PhD thesis, Heidelberg, Univ., Diss., 2014, 2014. Zsfassungen in dt. und engl. Sprache.

[185] WALLIKEWITZ, BODO H., DIRK HERTEL and KLAUS MEERHOLZ: *Cross-Linkable Polyspirobifluorenes: A Material Class Featuring Good OLED Performance and Low Amplified Spontaneous Emission Thresholds.* Chemistry of Materials, 21(13):2912–2919, July 2009.

[186] BAYERL, MICHAEL S., THOMAS BRAIG, OSKAR NUYKEN, DAVID C. MÜLLER, MARKUS GROSS and KLAUS MEERHOLZ: *Crosslinkable hole-transport materials for preparation of multilayer organic light emitting devices by spin-coating.* Macromolecular Rapid Communications, 20(4):224–228, April 1999.

Bibliography

[187] KÖHNEN, ANNE, NINA RIEGEL, JONAS H.-W. M. KREMER, HANS LADEMANN, DAVID C. MÜLLER and KLAUS MEERHOLZ: *The Simple Way to Solution-Processed Multilayer OLEDs – Layered Block-Copolymer Networks by Living Cationic Polymerization.* Advanced Materials, 21(8):879–884, February 2009.

[188] HERWIG, PETER T. and KLAUS MÜLLEN: *A Soluble Pentacene Precursor: Synthesis, Solid-State Conversion into Pentacene and Application in a Field-Effect Transistor.* Advanced Materials, 11(6):480–483, April 1999.

[189] ZAMBOUNIS, J. S., Z. HAO and A. IQBAL: *Latent pigments activated by heat.* Nature, 388(6638):131–132, July 1997.

[190] CHEN, TERESA L., JOHN JUN-AN CHEN, LUIS CATANE and BIWU MA: *Fully solution processed p-i-n organic solar cells with an industrial pigment – Quinacridone.* Organic Electronics, 12(7):1126–1131, July 2011.

[191] GLOWACKI, ERIC DANIEL, GUNDULA VOSS, KADIR DEMIRAK, MAREK HAVLICEK, NEVSAL SÜNGER, AYSU CEREN OKUR, UWE MONKOWIUS, JACEK GASIOROWSKI, LUCIA LEONAT and NIYAZI SERDAR SARICIFTCI: *A facile protection–deprotection route for obtaining indigo pigments as thin films and their applications in organic bulk heterojunctions.* Chemical Communications, 49(54):6063–6065, June 2013.

[192] SUNA, YUKI, JUN-ICHI NISHIDA, YOSHIHIDE FUJISAKI and YOSHIRO YAMASHITA: *Ambipolar Behavior of Hydrogen-Bonded Diketopyrrolopyrrole–Thiophene Co-oligomers Formed from Their Soluble Precursors.* Organic Letters, 14(13):3356–3359, July 2012.

[193] EDWARDS, JOHN H. and W. JAMES FEAST: *A new synthesis of poly(acetylene).* Polymer, 21(6):595–596, June 1980.

[194] BOTT, D. C., C. S. BROWN, C. K. CHAI, N. S. WALKER, W. J. FEAST, P. J. S. FOOT, P. D. CALVERT, N. C. BILLINGHAM and

R. H. FRIEND: *Durham poly acetylene: preparation and properties of the unoriented material.* Synthetic Metals, 14(4):245–269, May 1986.

[195] FEAST, W. J., J. TSIBOUKLIS, K. L. POUWER, L. GROENENDAAL and E. W. MEIJER: *Synthesis, processing and material properties of conjugated polymers.* Polymer, 37(22):5017–5047, October 1996.

[196] HAN, XU, XIWEN CHEN, GEORGE VAMVOUNIS and STEVEN HOLDCROFT: *Synthesis, Solid-Phase Reaction, Optical Properties, and Patterning of Luminescent Polyfluorenes.* Macromolecules, 38(4):1114–1122, February 2005.

[197] LIU, JINSONG, EKATERINA N. KADNIKOVA, YUXIANG LIU, MICHAEL D. MCGEHEE and JEAN M. J. FRÉCHET: *Polythiophene containing thermally removable solubilizing groups enhances the interface and the performance of polymer-titania hybrid solar cells.* Journal of the American Chemical Society, 126(31):9486–9487, August 2004.

[198] BJERRING, M., J.S. NIELSEN, N.C. NIELSEN, FREDERIK C KREBS, M. BJERRING, J.S. NIELSEN, N.C. NIELSEN and FREDERIK C KREBS: *Polythiophene by solution processing.* Macromolecules, 40:6012–6013, 2007.

[199] GEVORGYAN, SUREN, FREDERIK C KREBS, SUREN GEVORGYAN and FREDERIK C KREBS: *Bulk heterojunctions based on native polythiophene.* Chemistry of Materials, 20(13):4386–4390, 2008.

[200] KREBS, FREDERIK C, KION NORRMAN, FREDERIK C KREBS and KION NORRMAN: *Using Light-Induced Thermocleavage in a Roll-to-Roll Process for Polymer Solar Cells.* A C S Applied Materials and Interfaces, 2(3):877–887, 2010.

[201] *Method of thermocleaving a polymer layer.* US-Klassifikation 438/82, 257/E51.012; Internationale Klassifikation H01L51/48; Unternehmensklassifikation Y02E10/549, C08G2261/3223, H01L51/0004, H01L51/42, C08G2261/80, H01L51/0026, C08G2261/124, G03F7/2057, B41M5/36, C08G2261/364, H01L51/0036, B23K26/4065, G03F7/2012, C08G61/123, C08G61/126, C08G2261/1426, C08G2261/92,

Bibliography

C08G2261/3246, C08L65/00, H01L51/0038; Europäische Klassifikation C08G61/12D1, C08G61/12D1F, G03F7/20S3, G03F7/20A5, H01L51/00A2B2, B23K26/40B7H.

[202] LEE, JUNGHOON, A-REUM HAN, JAYEON HONG, JUNG HWA SEO, JOON HAK OH and CHANGDUK YANG: *Inversion of Dominant Polarity in Ambipolar Polydiketopyrrolopyrrole with Thermally Removable Groups.* Advanced Functional Materials, 22(19):4128–4138, October 2012.

[203] SALLEO, A., M. L. CHABINYC, M. S. YANG and R. A. STREET: *Polymer thin-film transistors with chemically modified dielectric interfaces.* Applied Physics Letters, 81(23):4383–4385, December 2002.

[204] SHERAW, C.D., T.N. JACKSON, D.L. EATON and J.E. ANTHONY: *Functionalized Pentacene Active Layer Organic Thin-Film Transistors.* Advanced Materials, 15(23):2009–2011, December 2003.

[205] SUBRAMANIAN, VIVEK, P.C. CHANG, J.B. LEE, S.E. MOLESA and S.K. VOLKMAN: *Printed organic transistors for ultra-low-cost RFID applications.* IEEE Transactions on Components and Packaging Technologies, 28(4):742–747, December 2005.

[206] KIM, SE HYUN, DANBI CHOI, DAE SUNG CHUNG, CHANWOO YANG, JAEYOUNG JANG, CHAN EON PARK and SANG-HEE KO PARK: *High-performance solution-processed triisopropylsilylethynyl pentacene transistors and inverters fabricated by using the selective self-organization technique.* Applied Physics Letters, 93(11):113306, September 2008.

[207] BOCK, C., D. V. PHAM, U. KUNZE, D. KÄFER, G. WITTE and CH WÖLL: *Improved morphology and charge carrier injection in pentacene field-effect transistors with thiol-treated electrodes.* Journal of Applied Physics, 100(11):114517, December 2006.

A. Appendix

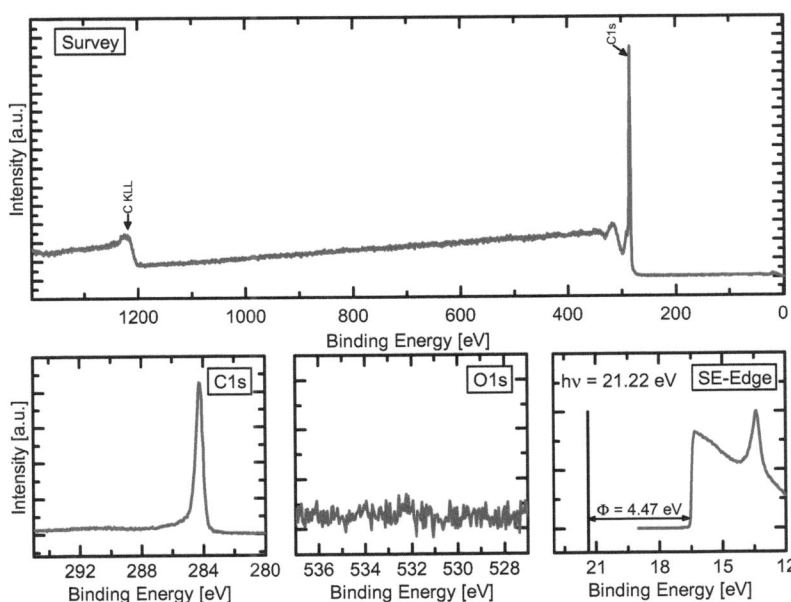

Figure A.1: XPS and UPS measurements of freshly cleaved HOPG substrate. Top: survey spectra which confirms the cleanliness of the used substrate. Bottom: detail spectra of C1s and O1s core levels and SE cutoff with the corresponding WF. The WF value of $\phi = 4.47\,\text{eV}$ is used to calculate the WFs of samples measured with KP.

A. Appendix

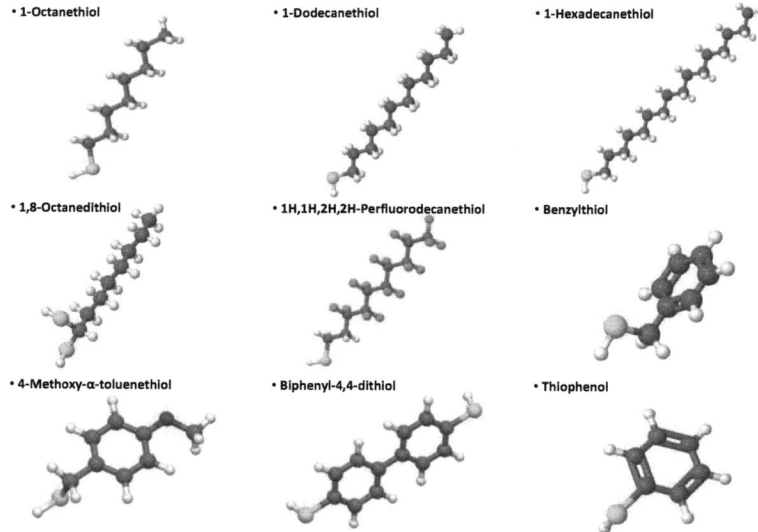

Figure A.2: Chemical structures of commercially available and in this work investigated SAM molecules (white: hydrogen, yellow: sulfur, grey: carbon, green: fluorine, red: oxygen).

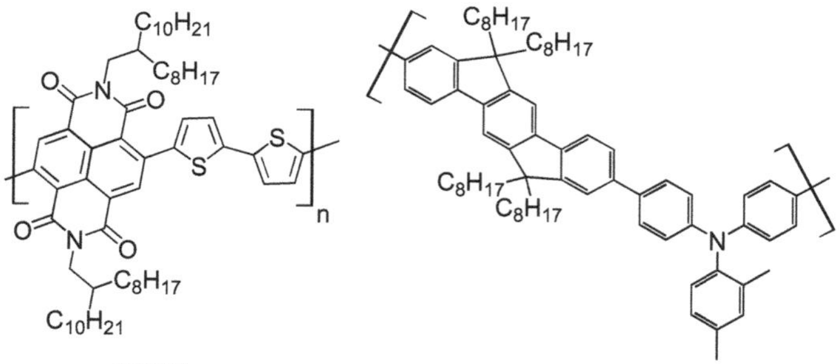

Figure A.3: Chemical structures of N2200 [174] and PIF8-TAA [162] semiconductors used in OFET devices.

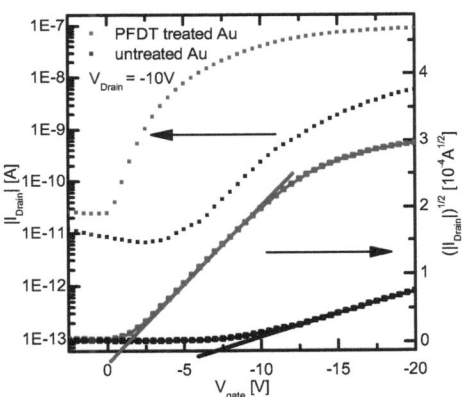

Figure A.4: OFET characteristics before and after treatment with PFDT SAM. Representative selection of data used to extract the V_{th} data presented Section 5.3.

A. Appendix

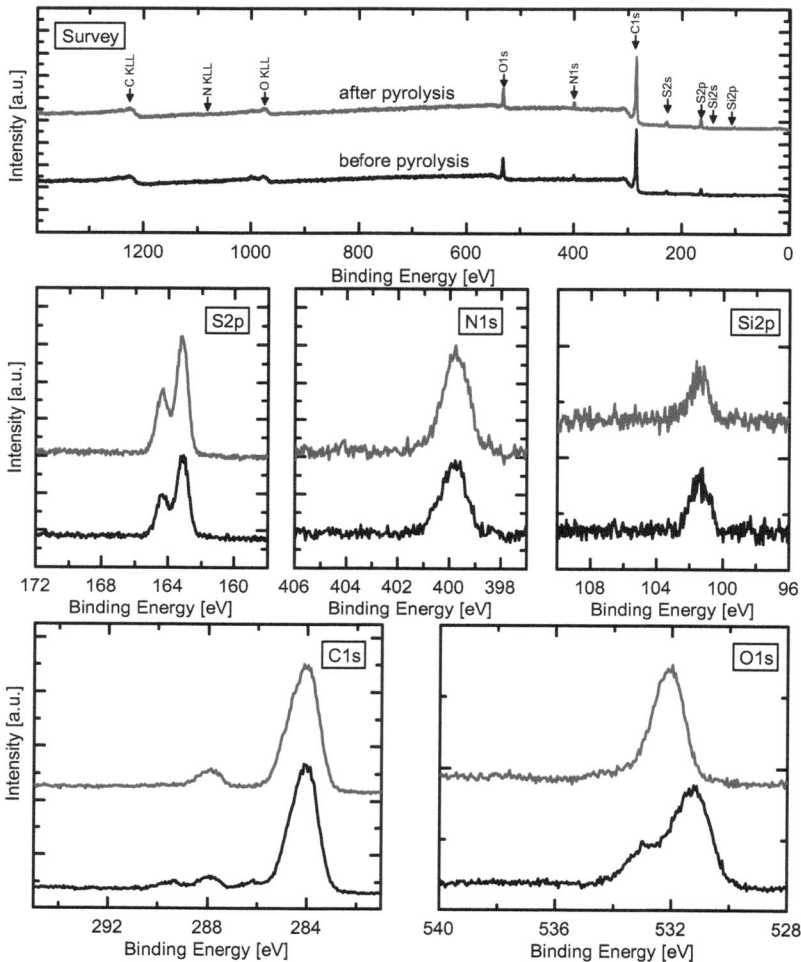

Figure A.5: Full XPS data of P(tHC-NDI-4HT2).

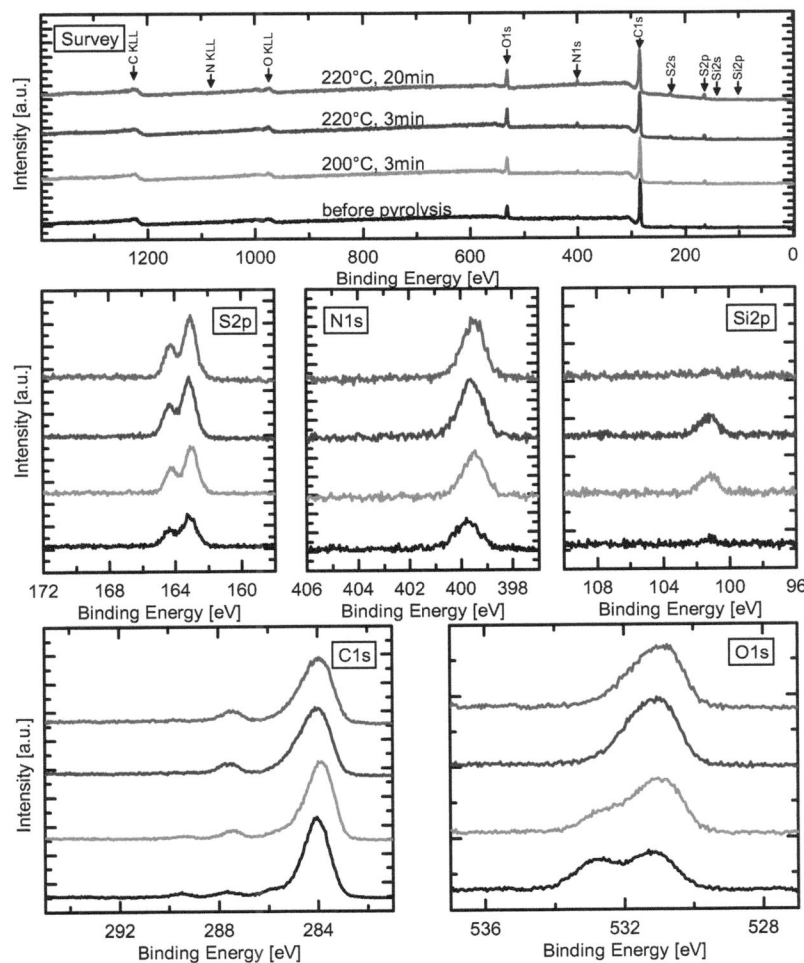

Figure A.6: Full XPS data of P(HtODC-NDI-T2).

A. Appendix

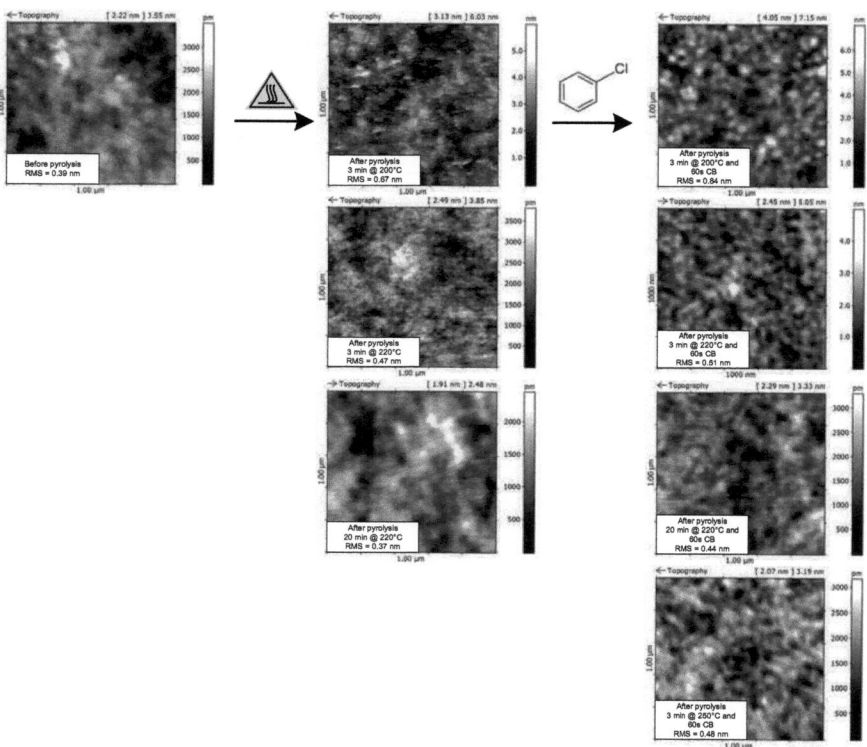

Figure A.7: All AFM measurements performed on P(HtODC-NDI-T2) polymer. Left: the spin coated layer, middle: P(HtODC-NDI-T2) polymer after different pyrolysis temperatures, right: samples after additional solvent treatment.

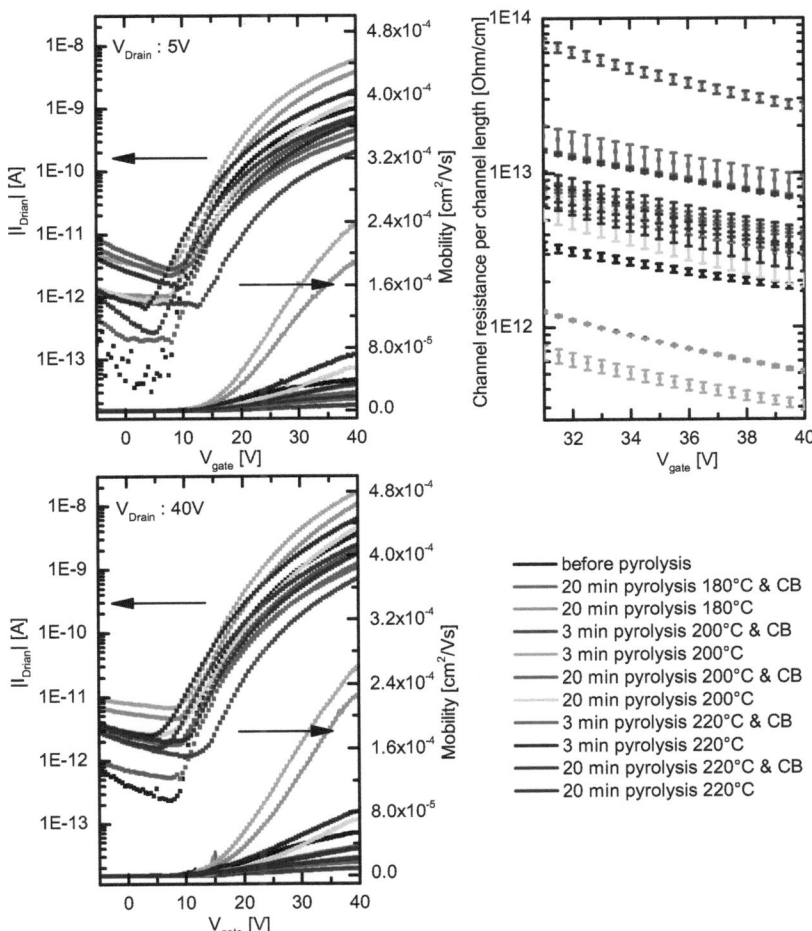

Figure A.8: OFET characteristics of P(HtODC-NDI-T2) used to extract the data shown in Figure 6.28.

Journal Publications, Patents, Conference Presentations and Supervised Theses

Journal Publications

1. Alt M., Schinke J., Hillebrandt S., Hänsel M., Hernandez-Sosa G., Mechau N., Glaser T., Mankel E., Hamburger M., Deing K., Jaegermann W., Pucci A., Kowalsky W., Lemmer U., Lovrinčić R., *Processing follows function: Pushing the formation of self-assembled monolayers to high throughput compatible timescales.* ACS Applied Materials & Interfaces.

2. Saive R., Scherer M., Mueller C., Daume D., Schinke J., Kroeger M., Kowalsky W., *Imaging the Electric Potential within Organic Solar Cells.* Advanced Functional Materials.

3. Saive R., Mueller C., Schinke J., Lovrinčić R., Kowalsky W., *Understanding S-shaped current-voltage characteristics of organic solar cells: direct measurement of potential distribution by scanning Kelvin probe.* Applied Physics Letters.

4. Stolz S., Scherer M., Mankel E., Lovrinčić R., Schinke J., Kowalsky W., Jaegermann W., Lemmer U., Mechau N., Hernandez-Sosa G., *Investigation of Solution-Processed Ultrathin Electron Injection Layers for Organic Light-Emitting Diodes.* ACS Applied Materials & Interfaces.

5. Jesper M., Schinke J., Alt M., Glaser T., Mankel E., Deing K., Jaegermann W., Pucci A., Kowalsky W., Lemmer U., Lovrinčić R., Hamburger M., *Dipolar SAMs for rational reduction of charge carrier injection barriers*

Journal Publications, Patents, Conference Presentations and Supervised Theses

in solution-processed organic thin-film devices. Langmuir - in preparation.

Patents

1. Title: *Thermische Abspaltung von löslichkeitsvermittelnden Gruppen bei Polymeren.* (declared but not yet accepted)

2. Title: *Austrittsarbeitsänderung durch selbst assoziierte Monolagen polarer Moleküle.* (declared but not yet accepted)

Conference Presentations (as a first author only)

1. J. Schinke, V. Rohnacher, M. Alt, S. Hillebrandt, M. Jesper, M. Hänsel, E. Mankel, T. Glaser, M. Hamburger, N. Mechau, R. Lovrinčić, A. Pucci, W. Jaegermann and W. Kowalsky; *Self-Assembled Monolayers;* Talk; Columbia University, Group of Prof. Xiaoyang Zhu; 2014; New York, USA.

2. J. Schinke, V. Rohnacher, M. Alt, S. Hillebrandt, M. Jesper, M. Hänsel, E. Mankel, T. Glaser, M. Hamburger, N. Mechau, R. Lovrinčić, A. Pucci, W. Jaegermann and W. Kowalsky; *A New Class of Self-Assembled Monolayers – towards an Ideal Interfacial Layer for Organic Electronic Devices;* Talk; MRS Spring Meeting 2014; San Francisco, USA.

3. J. Schinke, M. Alt, S. Hillebrandt, M. Hänsel, E. Mankel, T. Glaser, R. Lovrinčić, W. Jaegermann and W. Kowalsky; *Multi-analytical investigation of SAM formation on printing relevant timescales I: Kelvin probe and photoelectron spectroscopy;* Talk; DPG Dresden 2014.

4. J. Schinke, J. Heusser, M. Hänsel, J. Maibach, M. Alt, S. Stolz, W. Kowalsky, M. Kröger, N. Mechau E. Mankel, W. Jaegermann; *Improving the Contact Materials of Organic Electronic Devices: Polymeric Dipole Layers vs. Self Assembling Monolayers;* Talk; DPG Regensburg 2013.

5. J. Schinke, S. Hietzschold, R. Saive, L. Müller, M. Hamburger, M. Kröger, and W. Kowalsky; *Tuning the Surface Properties of Gold Electrodes in Organic Field-Effect Transistors Using Self-Assembled Monolayers;* Talk; DPG Berlin 2012.

6. J Schinke, D. Nanova, S. Hietzschold, M Scherer, C. Müller, R. Lovrinčić, W. Kowalsky; *Electronic and Morphological Studies of Organic Electronic Devices*; Poster; Hengstberger Symposium Heidelberg 2014.

7. J. Schinke, J. Heusser, M. Kröger, W. Kowalsky, E. Mankel, J. Maibach, W. Jaegermann; *Polymeric Dipole Layers vs. Self-Assembled Monolayers – an Ideal Interfacial Layer for Organic Electronic Devices;* Poster; ESPMI VII Israel 2013.

8. J. Schinke, J. Heusser, M. Kröger, W. Kowalsky, E. Mankel, J. Maibach, W. Jaegermann; *Polymeric Dipole Layers vs. Self-Assembled Monolayers – an Ideal Interfacial Layer for Organic Electronic Devices;* Poster; OIST Japan 2013.

9. J. Schinke, J. Heusser, M. Kröger, W. Kowalsky, E. Mankel, J. Maibach, W. Jaegermann; *Polymeric Dipole Layers vs. Self-Assembled Monolayers – an Ideal Interfacial Layer for Organic Electronic Devices;* Poster; MRS Fall Meeting 2012; Boston, USA.

10. J. Schinke, S. Hietzschold, R. Saive, L. Müller, M. Hamburger, M. Kröger, and W. Kowalsky; *Tuning the Surface Properties of Gold Electrodes in Organic Field-Effect Transistors Using Self-Assembled Monolayers;* Poster; MRS Fall Meeting 2011; Boston, USA.

Supervised Bachelor and Master Theses

- Sebastian Hietzschold (University of Heidelberg): *Optimierung von Grenzflächen in der organischen Elektronik durch selbstorganisierende Monolagen.* Bachelor thesis.

- Julian Heusser (University of Heidelberg): *Funktionalisierung von Metalloberflächen für organisch elektronische Bauteile.* Bachelor thesis.

Journal Publications, Patents, Conference Presentations and Supervised Theses

- Alexander Müller-Brand (University of Heidelberg): *Development of a Cryostat System and Its Subsequent Operation in Temperature-Dependent Conductivity Measurements on MoO_3-Doped CBP.* Master thesis.

- Marc Hänsel (University of Heidelberg): *Präparation und Charakterisierung selbstorganisierender Monolagen auf polykristallinen Substraten.* Master thesis.

Acknowledgement

Foremost I want to express my gratitude to my doctoral advisor Professor Wolfgang Kowalsky for his highly appreciated professional advice and his valuable feedback on arising problems. In addition I would like to thank him for introducing me to such an interesting topic and for letting me work with such great equipment at the InnovationLab. Furthermore, I thank him for giving me any freedom in the conduct of my project and for the support to travel to various conferences all over the world. My sincere thanks goes to my supervisors Dr. Michael Kröger and Dr. Robert Lovrinčić for sharing their broad knowledge with me, for their patience in introducing me to a subject area, that was not well known to me before and especially for their generous mentality. I wish you the best of success for your future projects. Also, I am grateful to all colleagues in Professor Kowalsky's research group for their efforts. I thank Professor Achim Enders for acting as second referee.

My thanks goes to Dr. Milan Alt in particular, who continuously offered his help and advice and devoted a lot of time during our common work on MORPHEUS project. Moreover I want to thank the whole iL team, especially Michaela Sauer for her friendly, fair and understanding support as well being such a cheerful soul. I also owe my gratitude to Kai Sudau, who always found a way to bring me to laugh. A big "Thank You" goes to the whole InnovationLab and especially the Analytics group for a wonderful time. We acted as an excellent team and we did a great job by setting up our institute. It is not possible to write all the names down, but Daniela Donhauser, Diana Nanova, Rebecca Saive, Michael Scherer, Tobias Glaser, Sebastian Beck, Sabina Hillebrandt, Sebastian "Pride" Stolz and Christian Müller must be mentioned! I thank all my Bachelor and Diploma students for their excellent work: Sebastian Hietzschold, Julian Heusser, Marc Hänsel, Alexander Müller-Brandt and Valentina Rohnacher. I am very grateful towards my dear friends and proof readers Dr. Richard Leys and Erdin Wegener for their effort in

Journal Publications, Patents, Conference Presentations and Supervised Theses

order to give me some great feedback, and especially to Dr. Robert Lovrinčić, who made many important remarks, while checking what I wrote. A very special thanks goes to our soccer group which gave me a lot of fun and also the possibility to get away from the daily grind.

Lastly I want to thank my family for being such a nice bunch towards me, especially my wife Małgorzata for her ongoing support throughout the past years.

Janusz

yes
I want morebooks!

Buy your books fast and straightforward online - at one of the world's fastest growing online book stores! Environmentally sound due to Print-on-Demand technologies.

Buy your books online at

www.get-morebooks.com

Kaufen Sie Ihre Bücher schnell und unkompliziert online – auf einer der am schnellsten wachsenden Buchhandelsplattformen weltweit! Dank Print-On-Demand umwelt- und ressourcenschonend produziert.

Bücher schneller online kaufen

www.morebooks.de

OmniScriptum Marketing DEU GmbH
Heinrich-Böcking-Str. 6-8
D - 66121 Saarbrücken
Telefax: +49 681 93 81 567-9

info@omniscriptum.com
www.omniscriptum.com

Printed by Books on Demand GmbH, Norderstedt / Germany